Mathematical Sciences Research Institute
Publications

7

Mathematical Sciences Research Institute
Publications

Alexandre J. Chorin Andrew J. Majda
Editors

Wave Motion: Theory, Modelling, and Computation

Proceedings of a Conference in Honor
of the 60th Birthday of Peter D. Lax

With 34 Illustrations

Springer-Verlag
New York Berlin Heidelberg London Paris Tokyo

Alexandre J. Chorin
Department of Mathematics
University of California
Berkeley, California 94720
USA

Andrew J. Majda
Department of Mathematics
Princeton University
Princeton, New Jersey 08544
USA

AMS Classification: 35LXX 35JXX 76F99 76H05 58GXX

Library of Congress Cataloging in Publication Data
Wave motion.
  (Mathematical Sciences Research Institute
  publications ; 7)
  Bibliography: p.
  Includes index.
  1. Wave motion, Theory of—Congresses.   2. Lax,
Peter D.—Congresses.   I. Lax, Peter D.   II. Chorin,
Alexandre Joel.   III. Majda, Andrew, 1949–
IV. Series.
QA927.W373   1987   531'.1133   87-23238

Text prepared in camera-ready form by the MSRI.
Printed and bound by R.R. Donnelley & Sons, Harrisonburg, Virginia.
Printed in the United States of America.

9 8 7 6 5 4 3 2 1

ISBN 0-387-96594-7 Springer-Verlag New York Berlin Heidelberg
ISBN 3-540-96594-7 Springer-Verlag Berlin Heidelberg New York

Peter D. Lax

Editors' Preface

The 60th birthday of Peter Lax was celebrated at Berkeley by a conference entitled "Wave motion: theory, applications, and computation" held at the mathematical Sciences Research Institute, June 9-12, 1986. Peter Lax has made deep and essential contributions to the topics described by the title of the conference, and has also contributed in important ways to many other mathematical subjects, and as a result this conference volume dedicated to him includes work on a variety of topics, not all clearly related to its title.

Peter Lax has made great contributions to mathematics and to science as a teacher, as a hero to emulate, and as a spokesman and advocate. He has, in the words of Lars Gårding, a genius for friendship. He is a famous wit. The large crowd of Peter Lax's friends, colleagues, coworkers, disciples and students that participated in the conference, buoyed by its admiration for Peter's achievements and by affection for his unique personality, was in a warm and cheerful mood and made this meeting into a genuine celebration.

The organizing committee would like to thank the staff of the Mathematical Sciences Research Institute, and particularly its director, Prof. Irving Kaplansky, for helping to organize this conference; it would like to thank the Mathematical Sciences Research Institute and the Lawrence Berkeley Laboratory for their sponsorship, and the Office of Energy Research, Office of Basic Energy Sciences, Engineering, Mathematical and Geosciences Division of the U.S. Department of Energy for providing financial support under contract DE-AC03-76SF0098.

Alexandre J. Chorin
Andrew J. Majda

TABLE OF CONTENTS

The following talks were given at the conference but have not been recorded:

C. Foias   "Conservation laws in wave number space and turbulence"

M. Kruskal   "Asymptotics beyond all orders"

A. Majda   "Concentration and oscillations in solutions of the incompressible fluid equations"

L. Nirenberg   "On non-linear second order elliptic equations"

LATTICE VORTEX MODELS AND TURBULENCE THEORY*

Alexandre Joel Chorin

Department of Mathematics
University of California, Berkeley, CA 94720

Abstract

Vortex lattice models for use in turbulence theory are explained, a simple one dimensional version is presented and is seen to provide reasonable qualitative information about intermittency and vortex stretching, and results obtained with three dimensional models are summarized.

* Work supported in part by the Director, Office of Energy Research, Office of Basic Energy Sciences, Engineering and Geosciences Division of the US Department of Energy, under contract DE-ACO3-765F00098.

Research supported in part by NSF Grant DMS-8120790.

Introduction

The lattice vortex models that we shall discuss are designed to provide qualitative understanding and quantitative predictions in turbulence theory.

The prospects for an accurate and detailed numerical solution of the Navier-Stokes or Euler equations in a fully turbulent situation are dim. Simple estimates of the number of degrees of freedom that must be resolved (see e.g. [2], [12]) show that the computation would be overwhelmingly large. The complexity of the phenomena observed in simplified problems (see e.g. [7], [8]) and estimates of the rate of growth of errors (see e.g. [4]) show that the problem is worse even than the simple counting argument indicates. Even if the computational problem were somehow solved, the problem of analyzing and understanding the results of the calculation would remain open. After all, nature presents us with detailed analogue calculations of turbulent flows and these have not yet led to a solution of the problem of turbulence.

There is a vast literature on the modeling of turbulence by averaged or phenomenological equations, usually involving empirical parameters. Such modeling is very useful as an engineering tool (see e.g. [2], [11]) but does not lead to general equations or to qualitative understanding [5]. Some of the reasons will be discussed below.

A way of tackling the turbulence problem is suggested by recent calculations ([7], [8]). The turbulent flow field is determined by its vorticity. The evolution of vorticity is typically dominated by a small number of coherent structures that interact through vortex / vortex interactions, through the exchange of vorticity (for a visualization, see [19]), and through processes of vortex stretching and creation. The details of these interactions are amazingly complex, but the calculations suggest that the overall statistics are independent of the details as long as certain constraints are satisfied. In many practical problems one is interested in scales of motion that are comparable with the scales of the coherent structures, and it is not desirable to carry out scaling or averaging operations that eliminate the spottiness (= "intermittency") of the solution. It is therefore

2

natural to turn to discrete models of the vorticity field; the lattice models are convenient because they allow one to separate scales in a natural fashion and keep the calculations relatively small.

What the lattice vortex models are not

At this point a disclaimer is appropriate. There are a number of mathematical constructions that use words similar to the ones we shall be using, and confusion is to be expected.

There is a substantial literature on lattice models and their hydrodynamical limits (see e.g. [18]) which is totally unrelated to the present work. As far as the present work is concerned, the Navier-Stokes and Euler equations have been derived, and the goal is to construct discrete models that mimic their behavior. If the models we introduce have a hydrodynamical limit it would presumably be, depending on the scaling, either the Navier-Stokes equations or some averaged model of turbulence.

Similarly, there have been in recent years a number of attempts ([13], [17]) to model continua by discrete models; even if these models can be made consistent with the equations, and even if they can be made more efficient than more standard methods (at present, a rather doubtful proposition), the problem of obtaining information about turbulence from solutions of the Navier-Stokes equations would remain as open as described in the introduction. The lattice models offered here and the lattice models proposed for example in the context of "cellular automata" are on opposite sides of the Euler or Navier-Stokes equations.

Vortex methods (see e.g. [8]) provide genuine approximations to solutions of the equations of motions and are thus quite distinct from the models discussed here. However, some of the features of the models were suggested by calculations with vortex methods. The closest analogues to the present models that I know of are the coarse grained collective variable models [1], [6], but the analogy is not close, in particular because the coarse-grained models do not include a vortex model of energy dissipation.

3

Euler's equations

The equations we shall be starting from are Euler's equations:

$$\partial_t \underline{\xi} + (\underline{u} \cdot \underline{\nabla})\underline{\xi} - (\underline{\xi} \cdot \underline{\nabla})\underline{u} = 0, \tag{1a}$$

$$\underline{\xi} = \text{curl } \underline{u}, \quad \text{div } \underline{u} = 0, \tag{1b,1c}$$

where $\underline{u}$ is the velocity, $\underline{\xi}$ is the vorticity, t is the time, and $\underline{\nabla}$ is the differentiation vector. These equations are appropriate for the analysis of scales larger than the dissipation scales, as discussed for example in [8]. For a derivation of these equations, see [10]; for a recent review of their theory, see [16]. The kinetic energy of a fluid whose motion obeys equations (1) is

$$T = \frac{1}{2} \int |\underline{u}|^2 d\underline{x} = \frac{1}{8\pi} \int d\underline{x} \int d\underline{x}' \frac{\underline{\xi}(\underline{x}) \cdot \underline{\xi}(\underline{x}')}{|\underline{x} - \underline{x}'|}, \tag{2}$$

where $\underline{x}$ is the position vector (see [15]). The mean squared vorticity is

$$Z = \int |\underline{\xi}|^2 d\underline{x}. \tag{3}$$

Equations (1) allow vortex tubes to stretch (for definitions, see [10]). Experiments, both numerical and physical, show that the stretching can be very substantial, and that as a result Z can become very large, possibly infinitely large in a finite time ([7], [16]).

Note that the integrand

$$B = B(\underline{x},\underline{x}') = \underline{\xi}(\underline{x}) \cdot \underline{\xi}(\underline{x}') / (8\pi |x-x'|)$$

in equation (2) is not positive definite. To see the significance of this fact, write

4

$$T = \int B d\underline{x} d\underline{x}' = \int_{|\,\underline{x}-\underline{x}\,'\,|\,<\,\epsilon} B d\underline{x} d\underline{x}' + \int_{|\,\underline{x}-\underline{x}\,'\,|\,\geqslant\,\epsilon} B d\underline{x} d\underline{x}' = T_< + T_> \,.$$

If ϵ is small and ξ is reasonably smooth, $T_< \geqslant 0$ and $T_>$ can change sign. As vortex tubes stretch and thin out, the vorticity associated with them increases and if ϵ is of the order of a tube diameter $T_<$ should increase. Since the sum T is non-increasing $T_>$ must decrease and possibly become negative. If $B < 0$, $\underline{\xi}(\underline{x}) \cdot \underline{\xi}(\underline{x}') < 0$, and the vortex tubes must fold. In picturesque language, $T_>$ provides the screen behind which $T_<$ can grow and vortex tubes can stretch.

Assumptions in the vortex lattice models

We now list the assumptions we shall be making in the lattice models. These assumptions are suggested by numerical experiment. The first four are merely global properties of Euler's equations:

(a)   the energy T is conserved as long as Z is finite. If Z is not finite, energy is non-increasing.

(b)   specific volume is conserved.

(c)   circulation ($\int_C \underline{u} \cdot d\underline{s}$, where C is any closed curve), is conserved.

(d)   the connectivity of vortex lines is invariant in three space dimensions; connectivity is important because it bestows some semblance of realism on the cascade models that lead to Kolmogorov's spectrum (see e.g. [9]).

(e)   vortex lines stretch rapidly and irreversibly.

Assumption (e) hides interesting mathematical difficulties, analogous to the problems in deriving the irreversible Boltzmann or Navier-Stokes

5

equations from the reversible Liouville equation. The solutions of Euler's equations are reversible at least as long as they remain smooth, yet the tendency towards stretching is very marked. Similarly, the solutions of Burgers' equation exhibit a tendency to form steep gradients for "most" initial data. Somehow the space of solutions and the "natural" space of initial data are different, but attempts to formulate this or similar statements precisely have not yet been successful. (For an interesting attempt, see [14].) Vortex stretching, subject to the conservation properties above, drives the flow towards turbulence. The details of the flow are hard to determine and can be viewed as random, with an unknown distribution. We shall pick that distribution at our convenience, in the hope that the choice does not unduly affect the gross statistical features of the flow.

These assumptions are similar to assumptions made by Childress [3] his Γ-model.

A one-dimensional model of vortex stretching

We now present a one dimensional "cartoon" of vortex stretching subject to energy conservation. In three dimensional space vortex tubes can move, stretch and fold; we mimic these processes on the line.

Consider the lattice $x_i = ih$, $i = 1,...,N$, $Nh = 1$, with periodic boundary conditions, where h is the length of the lattice bond. At time $t = 0$, place "vortices" of lengths $\ell_i = 1$ at the sites where i is odd and no vortices elsewhere. Endow the system with the energy

$$T = \sum_i \sum_{j \neq i} \text{sgn}(\ell_j) \cdot \text{sgn}(\ell_i) \frac{1}{r_{ij}} + \sum_i |\ell_i|, \qquad (4)$$

where $\text{sgn}(z) = 1$ if $z > 0$, -1 if $z < 0$, 0 if $z = 0$, and $r_{ij} = |(i-j)h| (\text{mod } \frac{1}{2})$. The double sum mimics $T_>$ with $\epsilon = h/2$. The

6

second term should mimic $T_<$ and depends on the unknown details of the vorticity distribution within each "vortex"; it is reasonable to set $|\ell_i|$ to be the contribution of each site to $T_<$ (and this statement may serve as a definition of ℓ_i). Let $T(t=0) = T_0$.

We now set up rules of motion for the vortices. Pick i at random, $1 < i \leqslant N$, and perform with equal probability one of the following operations:

(a) <u>Translation.</u> Consider the possibility of translating the "vortex" at $x_i = ih$ either to x_{i+1} or to x_{i-1}, if the new site is empty. If as a result $T > T_0$, do nothing; if $T \leqslant T_0$, perform the translation.

(b) <u>"Stretching".</u> When a real vortex is stretched its contribution to $T_<$ increases. In the one-dimensional model there is no realistic way to evaluate this increase, so we arbitrarily try to set $\ell_{i+1} = 2\ell_i$; if the resulting $T > T_0$ we do nothing, if $T \leqslant T_0$ the "stretching" is made. (A discussion of the scaling of $T_<$ in three dimension is given in [9]).

(c) <u>"Folding."</u> Consider setting $\ell_i = -\ell_i$; if the resulting $T \leqslant T_0$ do it, if $T > T_0$ do nothing.

T is thus allowed to decrease; it is plotted as a function of the number of moves in figure 1. T decreases when the "vortices" "fold" (i.e., one of a pair changes sign) and approach each other. T increases when "stretching" occurs. On a finite lattice the stretching eventually brings T back to T_0. When T reaches T_0 a new "vortex" configuration is obtained, stretched but with the same T as the original configuration. In figure 2, we plot part of the original configuration and the final configuration with the same T for $N = 40$. A negative ℓ_i is marked with an arrow pointing downward. We note that the new configuration is "intermittent", i.e., energy is not distributed uniformly in space. Intermittency is an important property of turbulence. Clearly without intermittency "vortex stretching" would be very limited. The formation of pairs of "vortices" of opposing sign corresponds to the formation of "hairpins" in real flow [8], [20].

7

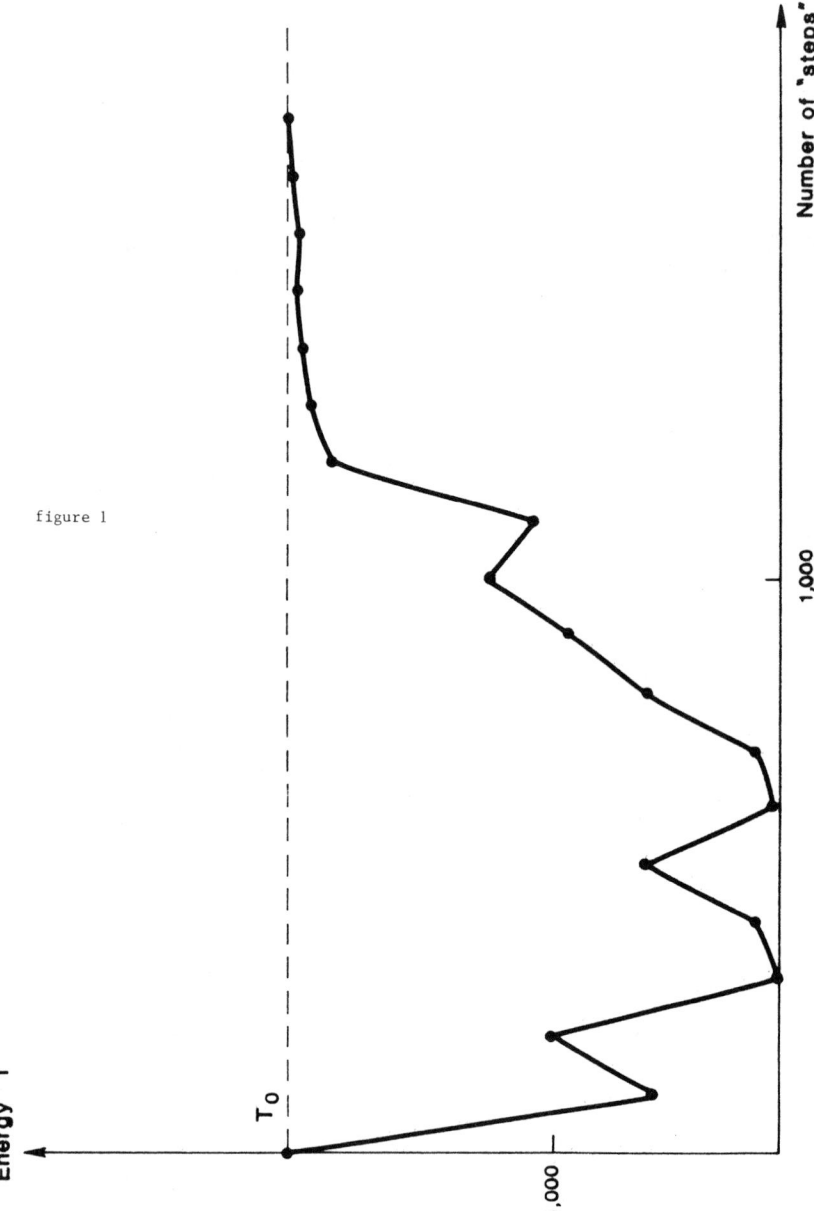

figure 1

8

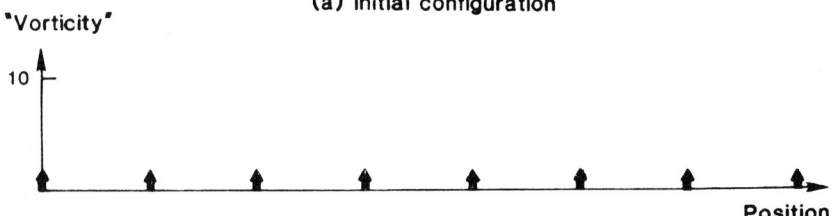

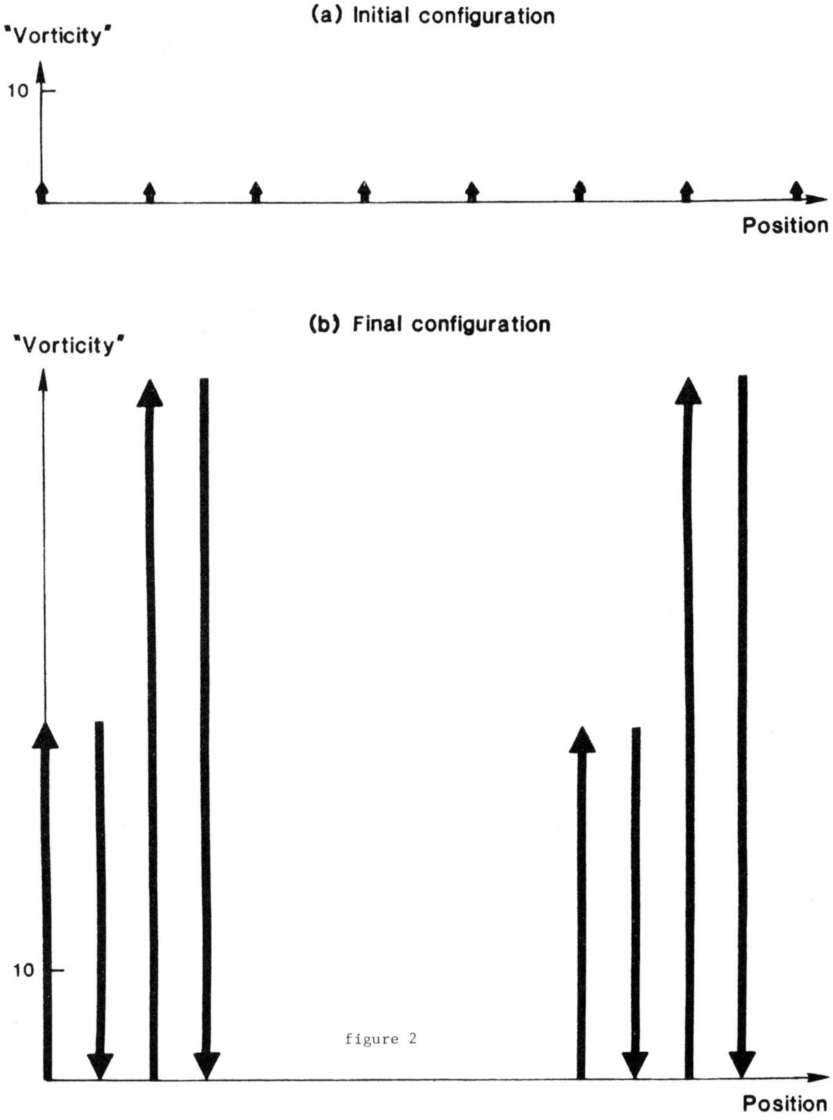

(a) Initial configuration

"Vorticity"

10

Position

(b) Final configuration

"Vorticity"

10

figure 2

Position

9

The length of the new "vortices" depends on h. Having gone as far as figure 2, we could refine the mesh and stretch some more. We could extract a part of the lattice, claim that its interactions with the rest of lattice are less significant than its internal dynamics, throw away the rest of lattice and concentrate on the part that remains. All these operations will be carried out in the three-dimensional case. Finally, if dissipation is important, on some scale one should allow cancellations between "vortices" of opposing sign. It is a reasonable conjecture that such cancellations are the major mechanism of enhanced energy dissipation in turbulent flow.

The simple model of this section is sufficient to exhibit intermittency, its relation to vortex stretching, hairpin formation and a mechanism of turbulent energy dissipation.

Summary of results in a three dimensional calculations

A three-dimensional calculation, of which the one-dimensional calculation of the preceeding section is a very simplified version, is presented in [9]. The vorticity field is assumed to consist of vortex tubes whose axes coincide with the bonds in a three-dimensional cubic lattice. An energy $T = T_> + T_<$ is defined, analogous to formula (2). The tubes are subjected to a random sequence of stretchings of which one is drawn in figure 3. The energy connected with the tubes before and after stretching is calculated in a manner consistent with equation (2), and with the constraints of conservation of specific volume and of circulation. Vortex tubes are not allowed to intersect nor to break. The energy decreases and then increases, in a manner similar to figure 1. When the energy returns to its initial value we view the outcome as an eddy that has broken down to smaller scales. Once an eddy has broken down, an eigth of the calculation is extracted, the mesh is refined and further breakdown becomes possible. This extraction/refinement process can be repeated ad infinitum. At each step in the process, the mean squared vorticity Z (equation 3) can be estimated.

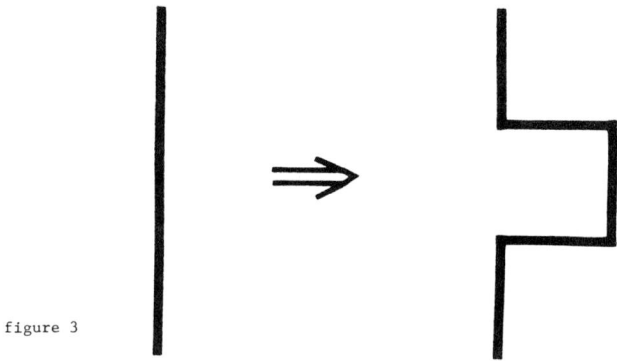

figure 3

Basic stretching in 3D

It is observed and proved that if the volume occupied by the "active" vorticity, i.e., vorticity still available for stretching, remains equal to the original volume, then the stretching will come to a halt, or else conservation of energy will be violated (as in the one dimensional model problem). The remedy is to force the stretching vorticity into ever smaller volumes (as in fact happens in solution of Euler's equation, [6], [7]). The sequence of decreasing volumes defines a similarity dimension D for the "active" portion of the flow. $T_>$ becomes negative and vorticity "tangles" appear. The amount of vortex stretching that occurs is a function of D. In homogeneous turbulence, the relation

$$Z(k) = k^2 E(k), \tag{5}$$

where k is the wave number and E, Z are respectively the energy and vorticity spectra (see e.g. [5]), determines the right increase in Z and leads to an equation for D that can be solved numerically, yielding an estimate $D \sim 2.4$. Equation (5) is a consequence of the definition of vorticity (1b) and of homogeneity. The calculation also yields an estimate of the inertial range exponent close to the Kolmogorov value $\gamma = 5/3$, a fact that is probably of little significance, since Kolmogorov's theory is dimensionally correct and any reasonably self-consistent model should reproduce it. It is significant however

11

that the Kolmogorov spectrum is not incompatible with intermittency, and that intermittency appears as a necessary consequence of vortex stretching and energy conservation.

There are two distinct length scales connected with the shrinking "active eddies". One is the cube root of the typical volume in which the active eddies are confined and is comparable with the radius of curvature of the vortex tubes; the other is the diameter of the vortex tubes. The second length scale decreases much more slowly than the first, as one can already see from figure 3 where the basic stretching is shown. The first length scale decreases as a result of this stretching by a factor $1/h \geqslant 2$, while the second decreases only by a factor $\sqrt{3} \sim 1.732$. The conclusion must be that as the energy cascade proceeds vortex tubes must divide into distinct subtubes that curve independently. A similar conclusion is reached in [3] by other arguments. This conclusion also constitutes a warning for straightforward vortex calculations in three dimensions: they must have a fine enough resolution to allow for the bursting of vortex tubes.

The time it takes an eddy to break down can be assumed to be comparable with its characteristic time $L/\sqrt{T}$, where L is its linear dimension. The sum of the characteristic times of the sequence of shrinking eddies converges, and thus, within the framework of the model, it takes a finite time t_* for Z to become infinite (see [7] for the details of the argument). To continue the calculation beyond t_* one has to allow for the interaction of several vorticity regions and for energy dissipation; these more elaborate models will be presented elsewhere.

Conclusion

On the basis of the results obtained so far, it is reasonable to predict that lattice vortex models will become a useful tool for investigating the structure of turbulence and a bridge between engineering models and the Euler and Navier-Stokes equations.

REFERENCES

[1] P.S. Bernard and B.B. Berger, A method for computing three dimensional turbulent flows, SIAM J. Appl. Math., 42, 453-470, (1982).

[2] T. Cebeci and A.M.O. Smith, Analysis of turbulent boundary layers, Academic, N.Y., (1974).

[3] S. Childress, A vortex tube model for eddies in the inertial range, Geophys. Astrophys. Fluid Dynamics, 29, 29-64, (1984).

[4] A.J. Chorin, The convergence of discrete approximations to the Navier-Stokes equations, Math. Comp., 23, 341-353, (1969).

[5] A.J. Chorin, Lectures on turbulence theory, Publish or Perish, Berkeley, (1975).

[6] A.J. Chorin, Crude numerical approximation of turbulent flow, in "Numerical solution of partial differential equations", Academic Press, 1976.

[7] A.J. Chorin, Estimates of intermittency, spectra and blow-up in developed turbulence, Comm. Pure Appl. Math., 24, 853-866, (1981).

[8] A.J. Chorin, The evolution of a turbulent vortex, Comm. Math. Phys., 83, 511-535, (1982).

[9] A.J. Chorin, Turbulence and vortex stretching on a lattice, Comm. Pure Appl. Math., in press, (1986).

[10] A.J. Chorin and J. Marsden, A mathematical introduction to fluid mechanics, Springer, N.Y., (1979).

[11] J. Ferziger, Higher level simulations of turbulent flow, Report TF-16, Dept. of Mech. Eng., Stanford (1981).

[12] C. Foias, O. Manley and R. Temam, Self-similar invariant families of turbulent flows, manuscript, 1986.

[13] U. Frisch, B. Hasslacher and Y. Pomeau, A lattice gas automaton for the Navier-Stokes equations, to appear in Phys. Rev. Lett., (1986).

[14] H. Glaz, Statistical behavior and coherent structures in two dimensional inviscid turbulence, SIAM J. Appl. Math., 41, 459-479, (1981).

[15] H. Lamb, Hydrodynamics, Dover, (1952), p.217.

[16] A. Majda, Mathematical foundations of incompressible fluid flow, lecture notes, Math. Dept., Princeton, (1985).

[17] R.H. Miller and K.H. Prendergast, Stellar dynamics in a discrete phase space, Astrophys. J., 151, 699-710, (1968).

[18] C. Morrey, On the derivation of the equations of hydrodynamics from statistical mechanics, Comm. Pure Appl. Math., 8, 279-327, (1955).

[19] J. Sethian, coherent structures in turbulent flow, Manuscript, UC Berkeley Math. Dept., 1986.

[20] E. Siggia, Collapse and amplification of a vortex filament, Phys. Fluids, 28, 794-804, (1985).

THE CURVE SHORTENING FLOW

By

C.L. Epstein and Michael Gage

August, 1986

In honor of Peter Lax's 60th birthday.

ABSTRACT

This is an expository paper describing the recent progress in the study of the curve shortening equation

$$(0.1) \qquad X_t = kN.$$

Here X is an immersed curve in $\mathbb{R}^2$, k the geodesic curvature and N the unit normal vector. We review the work of Gage on isoperimetric inequalities, the work of Gage and Hamilton on the associated heat equation and the work of Epstein and Weinstein on the stable manifold theorem for immersed curves. Finally we include a new proof of the Bonnesen inequality and a proof that highly symmetric immersed curves flow under (0.1) to points.

Keywords: curve shortening, heat flow, isoperimetric inequalities, stable manifolds

---

Research supported in part by NSF Grant DMS-8120790.

§1. Introduction

In this note we present some recent work on a popular equation, popularly known as the curve shortening flow:

$$(1.1) \qquad\qquad X_t = kN.$$

Here X is an immersion of S^1 in a surface, N is the unit normal and k is the geodesic curvature. This is "curve shortening" in virtue of the fact that it is the gradient flow of the length functional, see §2. The original motivation for studying (1.1) was to find a new and perhaps more natural proof of the existence of closed geodesics on Riemannian manifolds. To date, no geodesics have been found but the effort has uncovered many interesting results.

The problem has been most thoroughly studied in the case of locally convex planar curves. In [G1-3] and [G-H] it is shown that a convex initial curve shrinks to a point in finite time and its shape is asymptotically circular. The arguments used in these papers run the gamut from isoperimetric inequalities to apriori estimates for non-linear P.D.E.'s.

In the case of locally convex, immersed curves much less is known, though in [Ep-We] a stable-unstable manifold theorem is obtained that describes the behavior of (1.1) in neighborhoods of homothetic solutions. For a given total curvature, greater than 2π, there are two or more such solutions. One consequence of this work is the negative result that there is no universal, asymptotic behavior for immersed curves. In [Ab-La] sufficient conditions are obtained for a curve to develop a singularity before shrinking to a point.

In the following pages we will state the main results in the planar case and outline the methods used to prove them. In §2 we give a quick review of the theory for general curves. In §3 we outline the proof that a convex initial curve shrinks under (1.1) to a point in finite time with asymptotically circular shape. In §4 we discuss the stable-unstable manifold theorem for immersed curves and in §5 we prove that an immersed, locally convex curve, with a sufficiently large symmetry group converges to a point in finite time, with asymptotically circular shape.

16

The details for §§2-3 may be found in [G1-3] and [G-H], those for §4 in [Ep-We] and [Ab-La]. The result in §5 is new though the method of proof is closely related to that in §§2-3. At the end is an appendix containing a new proof of the Bonnesen inequality found by the second author.

By the asymptotic shape of a sequence of curves we mean the shape of a limiting object obtained by rescaling so that every curve in the rescaled sequence either encloses a fixed area or is of a fixed length. A detailed study of the limiting "shape" requires a careful analysis of the limiting behavior of the rescaled curvature functions.

Beyond the work described here, several authors have examined analogous problems for hypersurfaces in manifolds evolving under a flow like:

$$X_t = F(k_1,...,k_n)N;$$

here X is an immersion of a hypersurface and $F(k_1,...,k_n)$ is a function of the principal curvatures. For several choices of F it has been shown that convex initial data in $\mathbb{R}^{n+1}$ converges to a point in finite time with an asymptotically spherical shape. The interested reader should consult [Tso], [Ha], [Ch].

The convex or locally convex case is easier to handle technically because one has convenient global representation formulae for convex bodies in terms of their Gauss images [see §3]. Also, the positivity of the principal curvatures implies many useful analytic estimates; that is to say one has a good "maximum principal". To treat the non-convex case of imbedded planar curves, Matt Grayson [Gr] applied the dictum of Friedrichs', oft quoted by Peter, "to destroy invariance". By using local coordinate representations he was able to show that an embedded curve becomes convex before developing a singularity. Theorem 3.12, the Gage-Hamilton result implies that it shrinks to a point. Unfortunately space limitations prevent us from surveying either the non-convex case or the higher dimensional analogues. We refer the interested reader to the cited literature.

§2. Results for Arbitrary Plane Curves

2.1   In this section we review what is known about the evolution of smooth immersed planar curves.   Details can be found in [G-3] and [G-H].

Let X: $S^1 \times [0,T) \longrightarrow \mathbb{R}^2$ be the position vector of a family of curves indexed by t.   We write X(u,t), where u is a fixed spatial parameter for the curve, not necessarily arclength.   In fact arclength is a rather poor choice of parameter as it cannot be independent of time.   If one changes the space parameter to u = u(w,t) this may introduce a tangential component to the deformation vector field in (1.1):

(2.1) $X_t = kN + \alpha T,$

here T is the unit tangent vector to X.   Such changes in the right hand side do not effect the "geometric" properties of the flow.   This is fully discussed in §2.3.

Let $v = |x_u|$, then the arclength differential is given by: ds = v(u,t)du.   For general initial data we have a short time existence theorem:

Theorem 2.1:   *Given any smoothly immersed curve* $X_0: S^1 \longrightarrow \mathbb{R}^2$ *there is an* $\epsilon > 0$ *and a family of curves* X: $S^1 \times [0,\epsilon) \longrightarrow \mathbb{R}^2$ *such that*

$$X_t = kN$$

$$X(\cdot,0) = X_0.$$

For the proof see [G-H] §2 and example 2.8 in this paper.

The remaining interest in the problem is therefore in determining the largest time interval, [0,T) on which the family of solution curves is defined and determining the asymptotic behavior as t tends to T.

2.2   The first clues to the long term behavior are given by

18

the evolution of the macroscopic, geometric quantities associated to the curve, length and area. For an immersed curve, γ the area is defined by:

$$A = \int_{\gamma} xdy = \iint_{\mathbb{R}^2} \lambda(x,y,\gamma)dx \wedge dy, \quad (2.4)$$

here $\lambda(x,y,\gamma)$ is the winding number of γ relative to the point (x,y).

If the family of curves satisfies the equation

$$X_t = W$$

then elementary calculus leads to the formulae:

(2.2)
$$L_t = -\int_0^L <W,kN>ds$$

(2.3)
$$A_t = -\int_0^L <W,N>ds,$$

where $< , >$ denotes the standard inner product in $\mathbb{R}^2$. From (2.2) we see that in order to shorten a curve most efficiently one should use the flow

(2.4)
$$X_t = kN.$$

For this evolution (2.2) and (2.3) become:

(2.5)
$$L_t = -\int_0^L k^2 ds$$

(2.6)
$$A_t = -\int_0^L kds = -2\pi\nu,$$

19

here ν is the rotation number of Υ. From (2.5) and the Cauchy Schwarz inequality we obtain:

$$1/2(L^2)_t = -L \int_0^L k^2 ds \leqslant -(\int |k| ds)^2$$

$$\leqslant -4\pi^2 \nu^2$$

From this it follows that all initial data for (2.4) either shrinks to a point in finite time or develops a singularity. If the curve is immersed and the initial area is negative, then the curve must develop a singularity: the absolute value of the area is increasing while the length is decreasing. At some finite time the Banchoff–Pohl isoperimetric inequality, which states:

$$L^2 \geqslant \iint_{\mathbb{R}^2} \lambda^2(x,y,\Upsilon) dx \wedge dy,$$

will be violated. Thus we've proved:

Proposition 2.1: *If a curve encloses a negative signed area then its evolution under (1.1) develops a singularity in a finite time.*

This result appears in [Ab-La] with a somewhat different proof.

Similar calculations show that the time derivative of the curvature of a curve evolving under (1.1) satisfies:

(2.7) $$k_t = 1/v(k_u/v)_u + k^3$$

or in short hand:

(2.8) $$k_t = \frac{\partial^2 k}{\partial s^2} + k^3.$$

20

The operator $\partial/\partial s$, differentiation with respect to arclength, is defined by:

$$(2.9) \qquad \partial f/\partial s = (1/v)(\partial f/\partial u).$$

It is important to remember that the partial derivative with respect to t is taken keeping the parameter u fixed. The operator $\partial/\partial s$ is <u>not</u> a partial derivative and this severely diminishes the utility of the very simple looking equation (2.8).

The next result asserts that embedded curves remain embedded under the curve shortening flow. Heuristically, this follows by analysing the geometry of the curve at the first moment of contact. The tangents at the point of contact will be parallel and the curvature of the "inner" curve will be at least that of the "outer" curves. From this a contradiction follows, for if we evolve the curve backwards in time we see that the curve must have had a self intersection at an earlier time. Fortunately we do not need to run the flow backwards as the maximum principle suffices to prove:

Theorem 2.3 [G-H]: *Let* $X: S^1 \times [0,T) \longrightarrow \mathbb{R}^2$ *represent a one parameter family of closed curves satisfying* (1.1). *If the curvature satisfies* $|k| < C < \infty$ *for* $t \in [0,T)$ *and if the initial curve* $X(\cdot,0)$ *is embedded then* $X(\cdot,t)$ *is embedded for* $t \in [0,T)$.

2.3   Tangential Components

We conclude this section with a discussion of the effects of tangential components on (1.1).

Proposition 2.4: *A family of curves* $X(u,t)$ *which solves*

$$X_t = \alpha T + \beta N$$

can be converted into the solution of

$$X_t = \bar{a}T + \beta N$$

for any continuous function $\bar{a}$ by changing the space parametrization of the original solution. β must be a geometric quantity such as curvature; that is, its definition must not depend on the parametrization of the curve.

In other words changing the tangential component of the velocity vector, X_t affects only the parametrization--not the geometric shapes of the curves. The appropriate choice of the tangential component, a can simplify the analysis of the curve's behavior.

Remark: It is immediately clear from (2.2) and (2.3) that the rates of change of length and area are unaffected by the tangential component.

Proof of Proposition: Given $X(u,t)$: $S^1 \times [0,T) \longrightarrow \mathbb{R}^2$ as the original family of curves, let $u = u(w,\tau)$ and $t = \tau$ with $\partial u/\partial w > 0$, a reparametrization. This means $\bar{X}(w,\tau) = X(u(w,\tau),\tau)$. For τ fixed u maps a circle to a circle; it will be convenient to think of u: $\mathbb{R} \times \mathbb{R} \longrightarrow \mathbb{R}$ as a periodic function with $u(w+a,\tau) = u(w,\tau) + b$.
We calculate that:

(2.10)
$$
\begin{aligned}
\frac{\partial \bar{X}}{\partial \tau}(w,\tau) &= \frac{\partial X}{\partial \tau}(u(w,\tau),\tau) \\
&= \left[\left|\frac{\partial x}{\partial u}\right| \frac{\partial u}{\partial \tau} + a(u,\tau) \right] T(u(w,\tau),\tau) \\
&\quad + \beta(u(w,\tau),\tau)N(u(w,\tau),\tau) \\
&= \bar{a}(w,\tau)\bar{T}(w,\tau) + \bar{\beta}(w,\tau)\bar{N}(w,\tau).
\end{aligned}
$$

Hence the new parametrization satisfies a differential equation in which only the tangential component of the velocity vector has been changed; one still has $\bar{\beta}(w,\tau) = \beta(u(w,\tau),\tau)$, so that β and $\bar{\beta}$ have the same values at the same point on the curve considered as a subset of $\mathbb{R}^2$.
To see that any function $\bar{a}(w,\tau)$ can be obtained by an

22

appropriate choice of the function u, we let

$$u(w,0) = u_0(w): [0,a] \longrightarrow [0,b]$$

be an arbitrary parametrization of the original curve with $u_0(0) = 0$, $u_0(a) = b$ and du_0/dw positive. Here b is the period of the original parametrization. For each fixed w solve the following O.D.E.:

(2.11)
$$\frac{\partial u}{\partial \tau}(w,\tau) = \frac{\bar{a}(w,\tau) - a(u,\tau)}{\left|\frac{\partial x}{\partial u}\right|}$$
$$= F(u(w,\tau),\tau).$$

Here $\bar{a}(w,\tau)$ is the prescribed function and a and $|X_u|$ are explicitly given in terms of the values of u and t by the original parametrization. The function $u(w,\tau)$ will be a smooth function of w if the defining functions, $|X_u|$, $\bar{a}$ and a are smooth.

Differentiating (2.11) with respect to w shows that $\partial u/\partial w$ satisfies a linear equation and is always positive since du_0/dw is; hence $u(w,\tau)$ is a regular parametrization. To check that u has the correct periodicity observe that

$$F(u(w,\tau)-b,\tau) = F(u(w,\tau),\tau)$$

and that $u(w+a,\tau) - u(w,\tau) - b$ satisfies the linear equation:

(2.12)
$$\frac{\partial}{\partial \tau}(u(w+a,\tau)-u(w,\tau)-b) = F(u(w+a,\tau)-b,\tau) - F(u(w,\tau),\tau)$$
$$= G(u(w+a,\tau),u(w,\tau),\tau) \cdot [u(w+a,\tau)-b-u(w,\tau)]$$

where

$$G(x,y,\tau) = \begin{cases} (F(y,\tau)-F(x,\tau))/(y-x) & \text{if } x \neq y \\ F_x(x,\tau) & \text{if } x = y. \end{cases}$$

$G(u(w+a,u),u(w,\tau),\tau)$ is a smooth function of τ, hence the uniqueness of the solution to (2.12) implies that $u(w+a,\tau) = u(w,\tau) + b$ for all τ

23

and w. ∎

For later use we record:

Lemma 2.5: *If* $X_t = \alpha T + kN$ *then*

(2.13)
$$v_t = \alpha_u - k^2 v$$

and

(2.14)
$$T_t = (1/v)k_u + \alpha k N$$

Proof: Differentiate the equation in the hypothesis with respect to u, using the Frenet equations $T_u = vkN$ and $N_u = -vkT$. Since u and t are coordinates we have $X_{tu} = X_{ut} = (vT)_t$. Comparing the T and N components of this equation proves (2.13) and (2.14). ∎

Here are some examples:

Example 2.6: If $X(u,t)$ satisfies (2.1) then setting $u = w - t$ one obtains a family of curves satisfying

$$X_t(w,t) = -v(w-t,t)T + kN$$

on the other hand solving the O.D.E.

$$\frac{\partial u}{\partial t} = \frac{-1}{v(u,t)}$$

with $u(w,0) = w$ will give a parametrization of the family of curves satisfying

$$X_t = -T + kN.$$

These evolutions correspond to a "uniform rotation" of the space coordinate u with respect to w.

Example 2.7: In studying the evolution of locally convex curves Abresch and Langer use a parametrization in which the space parameter on each curve has constant (but not unit!) speed. This means that $v_u \equiv 0$. To determine which tangential component corresponds to this situation we differentiate (2.13) with respect to u. Since $v_u \equiv 0$ and $v_{ut} \equiv 0$ we have:

$$\alpha_{uu} = (k^2)_u v$$

or in terms of arclength

$$\alpha_{ss} = (-k^2)_s.$$

Integrating this twice gives:

$$\alpha(\sigma) = -\int_0^\sigma k^2 ds + b$$

where b can be chosen to make the average value of α equal to zero.

Example 2.8: Choosing the correct tangential component α also facilitates the proof of the short term existence theorem. In [G-H] section 2 the symbol of equation (2.1) is explicitly calculated and found to be weakly parabolic--i.e. it has positive and zero eigenvalues. Using the Nash-Moser implicit function theorem one can still show the existence of short time solutions. A simpler approach is to solve the equation

$$X_t = kN + \frac{v_u}{v^2} T$$

which can be written as

$$X_t = \frac{1}{v} \left[\frac{1}{v} X_u \right]_u + \frac{v_u}{v^3} X_u = \frac{1}{v^2} X_{uu}.$$

25

This modified equation is strictly parabolic and can be shown to have short time solutions using the standard implicit function theorem for Banach spaces. A change of variables yields a solution of (2.1). Also see [D] where this method was used to prove short term existence for the Ricci flow equation.

§3. Locally Convex Curves

3.1 As stated in the introduction, the locally convex case is somewhat easier because the "Gauss map" leads to global representation formulae. The Gauss map carries a planar curve to the unit circle by associating to a point on the curve the unit tangent at that point. One can parametrize the Gauss image by the angle θ between the tangent vector and the x-axis, see Figure 1.

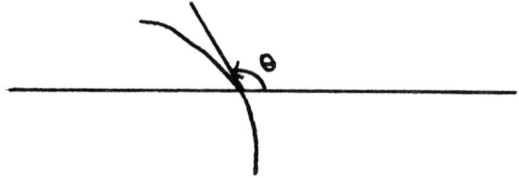

Figure 1

The geodesic curvature is defined by

$$\frac{d\theta}{ds} = k,$$

where s is the arclength parametrization. A curve is locally convex if $k > 0$; in this case θ defines a monotone parameter for the curve. If the curve is convex then θ varies from 0 to 2π, if the curve is only locally convex and closed then θ varies between 0 and $2\pi\nu$, for some integer ν given by:

$$2\pi\nu = \int_0^L k\,ds,$$

ν is called the rotation number of the curve.

If as above we fix a parameter u, different from arclength with $v(u)du = ds$ then $k = (1/v)(d\theta/du)$.

As we wish to study the curve shortening flow relative to the

27

parameter θ, we need to determine the tangential component to add to (1.1) so that $T_t = 0$; note that $T(\theta) = (\cos \theta, \sin \theta)$. From equation (2.14) we see that:

$$\alpha = -k_u/kv = -k_\theta.$$

Hence the evolution:

(3.1)
$$X_t = -k_\theta T + kN$$

will be used in the remainder of the paper. As noted above this change in evolution will not affect the geometry of the curves. For an immersed, locally convex curve, of total curvature $2\pi\nu$ we denote the domain of definition θ by $S^1_{2\pi\nu}$, the circle of length $2\pi\nu$, or the ν covered circle.

To describe a curve in terms of its Gauss image it is convenient to employ the support function:

$$p(\theta) = -\langle X, (-\sin \theta, \cos \theta) \rangle.$$

This is the distance from the tangent line at X to the origin, see Figure 2.

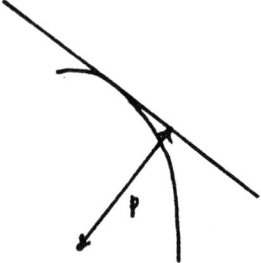

Figure 2

The support function evidently depends on the choice of origin; in applications one chooses the origin so as to reflect the geometric property under consideration, see Appendix. From the support function

28

the geometric invariants of the curve and their behavior under (3.1) are easily calculated:

Lemma 3.1: *The following formulae hold for locally convex curves:*

(3.2)
$$1/k = p + p_{\theta\theta}$$

(3.3)
$$L = \int_0^{2\pi\nu} p\,d\theta$$

(3.4)
$$2A = \int_0^{2\pi\nu} (p^2 - p_\theta^2)\,d\theta$$

We leave the proofs as exercises for the reader.

Lemma 3.2: *For the evolution equation* (3.1) *we have:*

(3.5)
$$p_t(\theta,t) = -k(\theta,t)$$

(3.6)
$$L_t = -\int_0^{2\pi\nu} k\,d\theta$$

(3.7)
$$A_t = -2\pi\nu$$

and

(3.8)
$$k_t = k^2 k_{\theta\theta} + k^3$$

Proof: Since $p = -\langle X,N\rangle$ and $N_t = 0$ for this evolution we have

$$p_t = -\langle -k_\theta T + kN, N\rangle = -k,$$

(3.6) now follows immediately from (3.3) and (3.5); (3.7) is proved by

29

differentiating (3.4) and using (3.5), integrating by parts and recalling formula (3.2); (3.8) is proved by differentiating (3.2). ∎

Notice the original curve can be reconstructed from p, using $p_\theta = <X,T>$ we have:

(3.9) $X = (-p \sin \theta + p_\theta \cos \theta, p \cos \theta + p_\theta \sin \theta)$.

To ensure that the curve produced is locally convex we require that $1/k = p + p_{\theta\theta} > 0$.

3.2   The Representation of Locally Convex Curves in Terms of Their Curvature

Up to translation we can recover the curve from the curvature as a function of θ, $k(\theta)$:

Lemma   3.3:   *A   positive   function $k: S^1_{2\pi\nu} \longrightarrow \mathbb{R}_+$ represents the curvature of a locally convex, closed curve if and only if:*

(3.10) $$\int_0^{2\pi\nu} \frac{e^{i\theta}}{k(\theta)} d\theta = 0$$

Remark: This is called the "closing condition" for curves with total curvature $2\pi\nu$.

Proof: Observe that, if the curve is parametrized by arclength then:

$$X(v) = \int_0^\sigma \frac{dX}{ds} ds = \int_0^\Sigma T ds$$

Changing variables from σ to θ we obtain:

30

$$(3.11) \qquad X(\theta) = \int_0^\theta (\cos \phi, \sin \phi) \frac{d\phi}{k(\phi)}.$$

Since $X(0) = X(2\pi\nu)$, (3.10) follows. On the other hand, given k satisfying (3.10), define a closed curve by (3.11). A brief calculation shows that this curve has curvature $k(\theta)$. ∎

Equation (3.8) and lemma 3.1 allow us to rephrase the curve shortening problem for locally convex curves as a parabolic P.D.E. for $k(\theta,t)$:

Proposition 3.4: *The problem of finding solutions to* $X_t = -k_\theta T + kN$ *is equivalent to finding a positive function* k: $S_{2\pi\nu}^1 \times [0,T) \longrightarrow \mathbb{R}_+$ *satisfying:*

$$(3.12) \qquad k_t = k^2 k_{\theta\theta} + k^3$$

$$(3.13) \qquad k(\theta,0) = \psi(\theta) > 0$$

where ψ *is a positive function on* $S_{2\pi\nu}^1$ *satisfying* (3.10).

Proof: If k satisfies (3.10), (3.12) and (3.13) then an easy calculation shows that

$$(3.14) \qquad \left[\int_0^{2\pi\nu} e^{i\theta}/k(\theta,t)d\theta \right]_t = 0$$

Thus (3.10) holds for each t and $k(\theta,t)$ represents, via (3.11) a closed locally convex curve. It is easily checked that this family of curves satisfies (3.1). On the other hand if the curves satisfy (3.1) and the initial curve is locally convex then lemma 3.2 and 3.3 show that the curvature will satisfy (3.12). An elementary application of the maximum principle shows that k remains positive under (3.12), see [G-H]. ∎

Remark: It is possible to prove short time existence and uniqueness theorems for equation (3.12) from standard results on parabolic equations. This avoids, for the special case of locally convex curves, the Nash–Moser machinery used in the proof of Theorem 2.1.

3.3   The Final Shape for Embedded Convex Curves

Since the derivative of area for an embedded curve is -2π, it is clear that the area will decrease to zero in a finite time and thus there is no possibility of obtaining a bounded solution to (3.12)-(3.13) or (3.1) for all time. In this section we will assume that the initial curve is an embedded convex curve ($\nu = 1$) and that a solution exists on the time interval $[0,T)$. Here $T = A_0/2\pi$ where A_0 is the (positive) area enclosed by the initial curve. From 3.7 we have that $\lim_{t \to T} A(t) = 0$; under these assumptions we will show that the curve shrinks to a point with asymptotically circular shape:

Theorem 3.7: *If the curve shortening evolution exists until the area goes to zero then a convex embedded curve converges smoothly to a shrinking circle.*

The proof follows from several isoperimetric inequalities. The classical isoperimetric inequality states: for an embedded planar curve

$$(3.15) \qquad\qquad L^2/A \geq 4\pi,$$

where equality holds only for a circle. In the argument below we will show that this isoperimetric ratio decreases under (3.1).

If we let r_{in} denote the radius of the largest circle which can be inscribed in a curve, γ and r_{out} the radius of the smallest circumscribed circle which encloses γ then the Bonnesen inequality states:

$$(3.16) \qquad\qquad rL - A \geq \pi r^2 \text{ for } r \in [r_{in}, r_{out}].$$

This is essentially a quantitative version of the isoperimetric inequality. A consequence of (3.16) is the following estimate for the "isoperimetric deficit":

$$(3.17) \qquad L^2/A - 4\pi \geq \frac{\pi^2}{A} (r_{out} - r_{in})^2,$$

for the proof see Corollary A.4. Using the Bonnesen inequality we will show that the isoperimetric ratio actually decreases to 4π as the area tends to zero. A new proof of (3.16) due to the second author is presented in an appendix.

A simple calculation using (2.5) and (2.6) shows that:

$$(3.18) \qquad (L^2/A)_t = \frac{-2L}{A} \left[\int_0^L k^2 ds - \pi L/A \right]$$

Proposition 3.8 [G1]: *For convex curves:*

$$\int_0^L k^2 ds - \pi L/A \geq 0.$$

Hence L^2/A decreases for convex curves under the curve shortening flow.

Proof: It is a consequence of the Bonnesen inequality that

$$(3.19) \qquad LA \geq \pi \int p^2 ds,$$

see the appendix. Applying the Schwarz inequality to (3.3) we obtain:

$$\begin{aligned} L^2 &= \left[\int_0^{2\pi} p(\theta) d\theta \right]^2 \\ &= \left[\int_0^L p(s) k ds \right]^2 \\ &\leq \left[\int_0^L p^2 ds \right] \left[\int k^2 ds \right]. \end{aligned}$$

33

Along with (3.19) this implies the assertion of the Proposition. ∎

In [G2] a stronger result is proved:

Proposition 3.9: *There exists a non-negative functional,* $F(\gamma)$ *defined for all convex,* C^1 *curves* γ *and which satisfies:*

(3.20)
$$\left[\int_0^L k^2 ds - \pi L/A\right] \geq \left[\int_0^L k^2 ds\right] F(\gamma).$$

Given a sequence of curves γ_i such that $\lim F(\gamma_i) = 0$, we consider the normalized curves $\eta_i = \sqrt{\dfrac{\pi}{A}}\, \gamma_i$; if these normalized curves lie in a fix bounded region of the plane, then the laminae, H_i which they enclose converge to the unit disk in the Hausdorff metric. Finally, $F(\gamma) = 0$ if and only if γ is a circle.

The functional F is given by:

$$F(\gamma) = (1 + \frac{\pi r_{in} r_{out}}{A} - \frac{2\pi(r_{in} + r_{out})}{L}).$$

To complete the proof of Theorem 3.7 we observe that

$$\lim_{t \to T} \inf L\left[\int_0^L k^2 ds - \pi L/A\right] = 0.$$

Were this not the case it would follow from (3.18) that L^2/A becomes negative before A goes to zero. Thus from (3.20) it follows that we can find a subsequence $\gamma_i = X(\cdot, t_i)$ with $t_i \longrightarrow T$ such that

$$\lim_{i \to \infty} F(\gamma_i) = 0.$$

This is so because:

34

$$L\int k^2 ds \geqslant \left(\int k ds\right)^2 = 4\pi^2.$$

Hence the subsequence of normalized curves η_i converges to a circle in the Hausdorff metric which in turn forces the isoperimetric ratio to tend to 4π along this subsequence. The quantity L^2/A is monotonely decreasing and therefore $\lim_{t \to T} L^2/A = 4\pi$. Now (3.17) implies that the shrinking curves, when normalized to enclose an area π, lie inside an arbitrarily thin annulus as $t \longrightarrow T$.

Remark: This is the weakest sense in which a family of curves can converge to a circle. One can show that the ratio of the maximum to minimum curvature of each curve converges to 1 and that k_θ (for the unnormalized curves) converges uniformly to zero, in spite of the fact that k tends uniformly to infinity, see [G-H].

3.4 Long Time Existence for Convex Curves

To complete the proof that a convex initial curve shrinks to a point we need to show that the solution can be continued so long as the area is positive. The principle task is to show that the curvature remains bounded--no cusps or corners develop.

To that end we introduce the median curvature k^*:

(3.21) $k^*(t) = \sup\{b \mid k(\theta,t) > b$ for θ in a interval of length $\pi\}$.

The following estimates for families of convex curves satisfying (3.1) are obtained in [G-H]:

Geometric Estimate: If $k^*(t)$ is given by (3.21) then:

$$k^*(t) \leqslant L(t)/A(t).$$

Integral Estimate: If $k^*(t)$ is bounded on [0,T) then

$$\int_0^{2\pi} \log k(\theta,t)d\theta$$

35

is bounded on [0,T) as well.

Pointwise Estimate: If $\int_0^{2\pi} \log k(\theta,t)d\theta$ is bounded on [0,T) then $k(\theta,t)$ and all its derivatives are bounded on [0,T).

Remark: The integral **and** pointwise estimates hold for any locally convex closed curves; they say in effect that the curvature cannot blow up on a small set, geometrically this means the curves cannot develop corners. Cusps are still a possibility.

Combining these estimates one proves:

Theorem 3.10: *If* $k: S^1 \times [0,T) \longrightarrow \mathbb{R}$ *and the area enclosed by the associated curves is bounded away from zero, then the curvature* k *is uniformly bounded on* $S^1 \times [0,T)$.

Proof: The lengths of the curves decrease during the evolution, so a uniform lower bound on area yields a uniform upper bound on $k^*(t)$ and, as a consequence of the integral and pointwise estimates, a uniform upper bound on $k(\theta,t)$.

The geometric estimate is quite easy:

Proof of Geometric Estimate: If $M < k^*(t)$ then $k(\theta,t) > M$ on some interval $[a,\pi+a]$. This implies that the convex curve lies between parallel lines whose distance is given by:

$$\int_a^{a+\pi} \frac{\sin(\theta-a)}{k(\theta,t)}d\theta \leq \frac{2}{M}.$$

36

The diameter is bounded by L/2 and the area is bounded by the "width" times the diameter. Since M can be chosen arbitrarily close to $k^*(t)$ it follows that:

$$k^*(t) \leqslant L/A.$$

The integral estimate is proved by dividing S^1 into intervals where $k > k^*$ and their complement. The Wirtinger inequality is used to obtain an appropriate bound on the intervals of large curvature, on the complement the curvature is bounded. See [G–H] pages 82–83 for details.

To obtain the pointwise estimate from the integral estimate we use a "reverse Poincaré inequality" true for solutions of (3.12):

Lemma 3.11: *There exists a constant D such that*

$$\int_0^{2\pi} \left(\frac{\partial k}{\partial \theta}\right)^2 d\theta \leqslant \int_0^{2\pi} k^2 d\theta + D$$

holds for $0 \leqslant t < T_0$.

Proof: We have:

$$
\begin{aligned}
\partial_t \int [k^2 - k_\theta^2] d\theta &= 2 \int (kk_t - k_\theta k_{\theta t}) d\theta \\
&= 2 \int (k_{\theta\theta} + k) k_t \, d\theta \\
&= 2 \int k^2 (k_{\theta\theta} + k)^2 d\theta \\
&\geqslant 0.
\end{aligned}
$$

Integrating this inequality completes the proof. ■

Further integral inequalities yield bounds on the higher derivatives of k. This allows one to continue the solution and we have therefore obtained:

Theorem 3.12: The solution of (3.12)–(3.13) continues until the area goes to zero and therefore convex initial data shrinks under 3.1 to a point with asymptotically circular shape.

§4.   Immersed Curves

In this section we treat locally convex, immersed curves in the plane.   This is a purely two-dimensional problem as a classical result of Hadamard, Stoker, et al. states that a locally convex, immersed, complete hypersurface is the boundary of a convex domain in dimensions three and greater.

As discussed above the simplest invariant associated to such a curve is its total curvature:

$$2\pi\nu = \int_0^L k(s)ds.$$

If $\nu > 1$ then there are at least two homothetic solutions to (2.4) and thus there is no "universal" asymptotic behavior as in the case $\nu = 1$.   In fact the isoperimetric arguments used in §3 do not extend to $\nu > 1$ and we have found it more convenient to use a rescaled equation introduced by Gage and Hamilton.   They use this flow to study the behavior of the curves as they collapse to a point. The rescaled flow is:

(4.1) $$k_\tau = k^2(k_{\theta\theta} + k) - k.$$

Here τ is a rescaled time parameter running from 0 to infinity.   The $-k$ term in (4.1) acts a constraint which forces curves initially enclosing an area $\pi\nu$ to continue to do so as they evolve.   These curves have the same shape as the curves which follow flow (2.1), but their size is different.

The point of view that we take is that (4.1) is a flow defined on a submanifold in a space of functions contained in $L^2(S_\ell^1)$.   The length of the circle, ℓ is tailored to reflect geometric properties of the class of curves.   For instance to study curves with total curvature $2\pi\nu$ we study (4.1) on functions periodic with period $2\pi\nu$.   If instead we want to study curves with total curvature $2\pi\nu$, invariant

39

under a q-fold rotation group, then we study (4.1) on periodic functions with period $2\pi\nu/q$.

The condition that the curves be closed is:

(4.2)
$$\int_0^{2\pi\nu} \frac{e^{i\theta}}{k(\theta)} d\theta = 0;$$

the condition that they enclose an area $\pi\nu$ is:

(4.3)
$$\int_0^{2\pi\nu} \frac{\cos\theta}{k(\theta)} \int_0^{\theta} \frac{\sin\phi}{k(\phi)} d\phi d\theta = \pi\nu.$$

As stated above (4.2) and (4.3) are preserved by (4.1). In this light the previous sections were devoted to proving that 2π periodic positive functions satisfying (4.2) and (4.3) tend to the constant 1 as τ tends to infinity in (4.1).

A stationary point for (4.1) translates into a homothetic solution of (3.8), which in turn represents a curve shrinking to a point retaining its initial shape. Several such initial curves are shown below:

Figure 3

The first step is to study the stationary points of (4.1), that is solutions to

(4.4) $$k^2(k_{\theta\theta} + k) - k = 0.$$

Dividing (4.4) by k^3 and multiplying by k_θ we see that (4.4) can be rewritten:

(4.5) $$\frac{d}{d\theta}\left[k_\theta^2 + k^2 - \log k^2\right] = 0$$

Thus in addition to $k \equiv 1$, we have a family of stationary periodic solutions; $k(\theta,E)$ given by:

$$(4.6) \qquad\qquad k_\theta^2 + k^2 - \log k^2 = E^2; \ E \in [1,\infty).$$

If $k_+(E)$ and $k_-(E)$ denote the roots of

$$k^2 - \log k^2 = E^2$$

then the fundamental period of $k(\theta,E)$ is given by:

$$(4.7) \qquad P_0(E) = 2 \int_{k_-(E)}^{k_+(E)} dk/\sqrt{E^2 - k^2 + \log k^2}.$$

In [Ab-La] it is shown that $P_0(E)$ varies monotonely between $\sqrt{2\pi}$ and π as E varies between 1 and ∞. If ν and q are relatively prime integers, with

$$1/2 < \nu/q < 1/\sqrt{2}$$

then there is a unique energy $E_{\nu,q}$ so that

$$P_0(E_{\nu,q}) = 2\pi\nu/q.$$

Let $w_{\nu,q}(\theta)$ denote the corresponding solution of (4.6); if we continue $w_{\nu,q}$ periodically to $S^1_{2\pi\nu}$ then it is the curvature of a closed curve with total curvature $2\pi\nu$ and a q-fold rotational symmetry. We study the flow (4.1) near to $w_{\nu,q}$ by setting $k = \delta + w_{\nu,q}$. The perturbation δ satisfies:

$$\delta_\tau = L_{\nu,q}\delta + a_{\nu,q}\delta_{\theta\theta} + b_{\nu,q}(\delta)\delta$$

where

$$(4.8) \quad \begin{aligned} L_{\nu,q}\,\delta &= w^2_{\nu,q}\,\delta_{\theta\theta} + (1 + w^2_{\nu,q})\delta \\ a_{\nu,q}(\delta) &= \delta(\delta + 2w_{\nu,q}) \\ b_{\nu,q}(\delta) &= (2w_{\nu,q} + 1/w_{\nu,q} + \delta)\delta. \end{aligned}$$

In [Ep-We] it is shown that generally the non-linear flow in (4.8) behaves, near the critical point $w_{\nu,q}$ like the linear flow:

$$(4.9) \quad \delta_T = L_{\nu,q}\,\delta.$$

The choice of underlying function space enters through its effect on the spectral theory of $L_{\nu,q}$.

In the sequel we will abbreviate $L_{\nu,q}$ to L and $w_{\nu,q}$ to w, where it will not cause confusion.

Generally L is a formally self adjoint operator relative to the positive measure $d\theta/w^2$. The spectral theory is greatly facilitated by the facts that:

$$(4.10) \quad \begin{aligned} &\text{a)} \quad Lw = 2w \\ &\text{b)} \quad L\partial_\theta w = 0 \end{aligned}$$

As w is a positive function it follows that the largest positive eigenvalue is 2. The function $\partial_\theta w$ has exactly two zeros on $S^1_{2\pi\nu/q}$. From elementary Stum-Lionville theory one concludes:

Theorem 4.1 [Ep-We]: *On $L^2(S^1_{2\pi n\nu/q}, d\theta/w^2)$ L is a self adjoint operator with $2n - 1$ positive eigenvalues, a one dimensional null space and the remaining eigenvalues negative tending to $-\infty$.*

The one dimensionality of the null space follows from the monotonicity of $P_0(E)$. The sense in which the linear theory models the nonlinear flow is given in the following "stable-unstable" manifold theorem:

Theorem 4.2 [Ep-We]: *Let n divide q*

I.   *In the space of* $2\pi n\nu/q$-*periodic functions there is a codimension 2n submanifold of data near to* $w_{\nu,q}$ *such that the solution to (4.1) with initial data in this set tends to* $w_{\nu,q}$ *as* $\tau \longrightarrow \infty$. *All the data in this set represents closed curves.*

II.   *There is a (2n - 1)-dimensional submanifold of initial data near to* $w_{\nu,q}$ *such that the solution to (4.1) with initial data in this set tends to* $w_{\nu,q}$ *as* $\tau \longrightarrow -\infty$. *If* n = q *then this set has a codimension 2 submanifold of data representing closed curves; if* n < q *then all the data represents closed curves.*

Remark: We assume that ν and q are relatively prime.

A quick count reveals that we are missing a dimension. This corresponds to the zero eigenspace and it is the so-called "center manifold". In our case this is very simple for equation (4.4) is homogeneous so all the functions $\{w_{\nu,q}(\theta + a): a \in [0,2\pi\nu/q)\}$ represent distinct stationary points in $L^2(S^1_{2\pi\nu})$. Geometrically this corresponds to rotating the curve in the plane. Hence the center manifold is a circle composed of stationary points.

It is now apparent that no theorem like Theorem 3.7 can hold for immersed curves. In the space of $2\pi\nu$-periodic data there isn't even a stationary point which attracts an open set in the space of data that satisfies (4.2) and (4.3).

After briefly describing the proof of Theorem 4.2 for the case n = q we will discuss the dynamics of (4.1) on several different function spaces.

The proof closely follows the proof of the stable manifold theorem for ordinary differential equations:

We introduce spectral projection operators π_+, π_0, π_- corresponding to the positive, null and negative eigenspaces of L, respectively. Then we rewrite (4.1) as an integral equation:

44

$$T(\delta_i \delta_0) = e^{L\tau}\pi_-\delta_0 + \int_0^\tau e^{L(\tau-\sigma)}\pi_-(a(\delta)\delta_{\theta\theta} + b(\delta)\delta)d\sigma$$

(4.11)
$$-\int_\tau^\infty \pi_0(a(\delta)\delta_{\theta\theta} + b(\delta)\delta)d\sigma$$

$$-\int_\tau^\infty e^{L(\tau-\sigma)}\pi_+(a(\delta)\delta_{\theta\theta} + b(\delta)\delta)d\sigma.$$

A solution of $T(\delta,\delta_0) = \delta$ is a solution to (4.1) with initial data given by:

$$(4.12) \quad \delta(0) = \pi_-\delta_0 - \int_0^\infty [\pi_0 + e^{-L\sigma}\pi_+](a(\delta)\delta_{\theta\theta} + b(\delta)\delta)d\sigma.$$

One thinks of T as an operator on $L^2(\mathbb{R}_+, C^\infty(S^1_{2\pi\nu})) \times C^\infty(S^1_{2\pi\nu})$. Next we cook up a norm that makes T into a bounded operator from a space to itself. For example one could use:

$$|f|^2_{\mu,s} = \int_0^\infty \|f(t)\|^2_s dt + \sup[\|f(t)\|^2_{s-2}e^{2\mu t} + \|f(t)\|^2_{c^{s-2}}];$$

Here $\|f\|_s$ is the norm on the Sobolev space $H^s(S^1_{2\pi\nu})$ and μ is a positive constant determined by the eigenvalues of L. Relative to a norm like $|\cdot|_{\mu,s}$, $s \geq 2$, $T(\delta,\delta_0)$ is a contraction, for small enough δ_0 and δ. Consequently there is a fixed point establishing the existence of the map (4.12) between stable data for L and stable data for (4.8).

The important thing to note is that the length of the underlying circle enters into the analysis of T only through the spectral decomposition of L. If, for instance, we work with $2\pi\nu/q$ periodic functions then L has a single positive eigenvalue and a one-dimensional null space. Thus the stable manifold near $w_{\nu,q}$ in the space of $2\pi\nu/q$-periodic functions is of codimension 2. The eigenspace corresponding to 2 is simply the normal vector to the

45

hypersurface of data enclosing an area $\pi\nu$, the null space is as above. Geometrically this means that every curve with total curvature $2\pi\nu$ and a q-fold symmetry, with curvature sufficiently near to $bw_{\nu,q}(\theta + a)$, for some a and b > 0 converges to a point with asymptotic shape given by $w_{\nu,q}$. In this space there is only one other critical point at $k \equiv 1$; it is also stable. Let $\mathcal{S}$ denote the data not attracted to either 1 or $\{w_{\nu,q}(\theta + a): a \in S^1_{2\pi\nu/q}\}$, the following questions about $\mathcal{S}$ seem natural:

1)   Are there any bounded trajectories in $\mathcal{S}$?

2)   Does $\mathcal{S}$ have non-empty interior?

As a final case we consider curves with a high degree of symmetry: A curve with total curvature $2\pi\nu$ and q-fold symmetry has a high degree of symmetry if $\nu/q < 1/2$. The only stationary point in the space of $2\pi\nu/q$-periodic data is the constant 1. The linear analysis and Theorem 4.1 imply that this is a stable critical point. It is natural to conjecture that all such data should converge under (3.1) to a point with asymptotic shape a ν-covered circle. In the next section we will outline a proof of this conjecture. The proof relies on the ideas in §§2-3 but is actually a little easier due to a stronger form of Bonnesen's inequality, (A.5).

46

§5. Highly Symmetric Curves

In this section we will show that a locally convex, closed, highly symmetric curve (see §4) tends under 3.1 to a point. The shape is asymptotically a ν-covered circle. The argument is similar to that presented in §§2-3, though a stronger Poincaré inequality, (A.5) leads to a simpler convergence proof. The argument requires a little geometric information about highly symmetric curves:

Lemma 5.1: *A highly symmetric, locally convex curve does not pass through its center of symmetry, nor does it have a self intersection within a period.*

Proof: Suppose γ is a curve with total curvature $2\pi\nu$ and a rotational symmetry of order q passing through its center of symmetry. We will show that

$$\nu/q > 1/2.$$

We assume that ν and q are relatively prime. The curve can be written as a union of petals

$$\gamma = \bigcup_{i=1}^{q} \gamma_i,$$

where γ_i is carried into γ_{i+1} by the rotation. A petal is defined by following the curve from its center of symmetry to its next passage through this point.

Figure 4

No open set on a given petal is mapped into the same petal by the rotation for an elementary connectivity argument would then show that the entire petal is fixed as a set by the rotation. From this we could conclude that the curve consists of a single petal, which is clearly impossible. As ν and q are relatively prime the same argument shows that, as an immersed curve the petal has (q - 1) distinct, disjoint images. The total curvature of a petal is at least π, hence we see that

$$2\pi\nu > q\pi$$

or

$$\nu/q > 1/2$$

as asserted. The last part of the argument also proves the second assertion of the lemma. ∎

If the curve is highly symmetric then circles with small radius about the center of symmetry lie entirely "inside" the curve; let r_{in} denote the radius of the largest such inscribed circle and r_{out} the radius of the smallest circumscribing circle about this center. Let p denote the support function relative to the center of symmetry. From their definitions it is evident that

(5.1) $r_{in} \leq p \leq r_{out}.$

For $r \in [r_{in}, r_{out}]$, p - r vanishes somewhere and the separation between the consecutive zeros of $(p(\theta) - r)$ is at most $2\pi\nu/q$. Thus we can apply corollary A.6 to conclude:

Lemma 5.2: *For highly symmetric, locally convex curves*

48

(5.2) $rL - A - \nu\pi r^2 \geqslant (1/2)[(q/2\nu)^2 - 1] \displaystyle\int_0^{2\pi\nu} (p-r)^2 d\theta,$

for $r \in [r_{in}, r_{out}]$.

In light of (5.1) we can insert $p(s)$ into (5.2) and integrate with respect to arclength to obtain:

(5.3) $LA - \pi\nu \displaystyle\int_0^L p^2(s)ds \geqslant 1/2[(q/2\nu)^2-1] \int_0^{2\pi\nu}\int_0^L (p(\theta)-p(s))^2 ds\, d\theta.$

From this it follows, as in Proposition 3.8 that if $L(t)$ and $A(t)$ are the length and area of a highly symmetric curve evolving under (3.1) then:

(5.4) $(L^2/A)_t = -2(L/A)\left[\displaystyle\int_0^L k^2 ds - \pi\nu L/A\right] < 0.$

Now suppose that $A(t)$ tends to zero as $t \longrightarrow T$; as above it follows that there is a subsequence $\{t_i\}$ tending to T such that $L(\int k^2 ds - \pi\nu L/A)$ tends to zero along this sequence, see §3.3. On the other hand letting $p(\cdot,t)$ denote the support functions relative to the center of symmetry we have:

(5.5) $L(\int k^2 ds - L\pi\nu/A) \geqslant \left[L^3\Big/\displaystyle\int_0^L p^2 ds - \pi\nu L^2/A\right] \geqslant 0.$

Therefore the quantity in the middle of (5.5) must tend to zero along the subsequence as well. We rewrite (5.3) as:

$$(5.6) \quad A/L^2 - \pi \nu \int_0^L (p/L)^2 \, ds/L$$

$$\geq 1/2 [(q/2\nu)^2 - 1] \int_0^{2\pi\nu L} \int_0^L \left[\frac{p(\theta)}{L} - \frac{p(s)}{L} \right]^2 \frac{ds}{L} d\theta.$$

If we replace s by $\sigma = s/L$ and p by $\tilde{p} = p/L$, then we have renormalized the curves so that they all have unit length. It follows from (5.5) and (5.6) that along the subsequence $\{t_i\}$

$$(5.7) \quad \lim_{i \to \infty} \int_0^L \int_0^{2\pi\nu} (\tilde{p}_i(\theta) - \tilde{p}_i(\sigma))^2 \, d\theta \, d\sigma = 0,$$

where $\tilde{p}_i = \tilde{p}(\cdot, t_i)$.

From the formulae for the enclosed area and length:

$$(5.8) \quad A/L^2 = 1/2 \int_0^{2\pi\nu} (\tilde{p}^2(\theta) - \tilde{p}_\theta^2(\theta)) \, d\theta$$

$$1 = \int_0^{2\pi\nu} \tilde{p}(\theta) \, d\theta$$

the fact that $\tilde{p}_i(\theta)$ is a $2\pi\nu/q$–periodic function and the Poincaré inequality it follows that $\{\tilde{p}_i(\theta)\}$ is a uniformly bounded and equicontinuous family. Thus by the Arzela–Ascoli theorem there is a further subsequence $\{\tilde{p}_{i_j}(\theta)\}$ tending uniformly to a continuous limit.

Hence an even further subsequence $\{\tilde{p}_{i_{j_k}}(\sigma)\}$ must converge almost everywhere. Let $\tilde{p}_*(\theta)$ and $\tilde{p}_*(\sigma)$ denote the respective limits. From (5.7) it follows that

$$(5.9) \quad \tilde{p}_*(\theta) = \tilde{p}_*(\sigma)$$

almost everywhere and therefore each side of (5.9) is constant. From

50

this and (5.8) it follows that $\hat{p}_*(\theta) = 1/2\pi\nu$.   From (5.6) it now follows that

(5.10) $\lim_{i\to\infty} L^2(t_i)/A(t_i) = 4\pi\nu.$

Since L^2/A is monotonely decreasing it follows that L^2/A tends to $4\pi\nu$ as $t \longrightarrow T$. From this and the corollary of Lemma 5.2

(5.11) $L^2/A - 4\pi\nu \geqslant \dfrac{\pi^2\nu^2}{A} (r_{out} - r_{in})^2$

we conclude that as A tends to zero the renormalized curve is contained in an arbitrarily thin annulus.   Note that (5.11) implies that

$$\lim_{t\to T} \hat{p}(\theta,t) = 1/2\pi\nu$$

uniformly and thus (5.8) implies that $\hat{p}_\theta$ tends to zero in L^2.

To complete the argument we need to show that the evolution (3.8) exists so long as the area enclosed is positive.   The argument is very similar to that given in §3:

We introduce the median curvature:

$k^*(t) = \sup\{b \mid k(\theta,t) > b$ for θ in an interval of length $\pi\}.$

As before an analytic argument using the Wirtinger inequality and Lemma 3.11 shows that so long as $k^*(t)$ is bounded, $k(\theta,t)$ and all its derivatives are also bounded.   From the maximum principle it follows that $k_{min}(t) = \min_\theta k(\theta,t)$ is non-decreasing and thus the solution can be continued as long as $k^*(t)$ is bounded.

The proof that $k^*(t)$ is bounded is quite easy for $k(\theta,t)$ is $2\pi\nu/q < \pi$-periodic and therefore

$$k^*(t) = k_{min}(t).$$

The length of the curve is given by:

51

$$L = \int_0^{2\pi\nu} d\theta/k(\theta,t) < \int_0^{2\pi\nu} d\theta/k_{min}(t) = 2\pi\nu/k_{min}(t).$$

Hence $k^*(t) \leqslant 2\pi\nu/L(t)$. As $L^2/A \geqslant 4\pi\nu$ we have $k^*(t) \leqslant 1/2(L(t)/A(t))$; $L(t)$ is decreasing and therefore so long as $A(t) > 0$, $k^*(t)$ is bounded. We can therefore continue the solution to 3.8. Altogether we've shown:

Theorem **5.3**: *Immersed, locally convex, closed curves with total curvature $2\pi\nu$ and a q-fold rotational symmetry with $\nu/q < 1/2$ tend to a point under (3.8) with asymptotic shape a ν-covered circle.*

The argument presented here only establishes convergence in a rather weak sense, however one can adapt the estimates of [G-H] to show that:

$$k_{max}(t)/k_{min}(t) \longrightarrow 1$$

and even that

$$k_\theta(\theta,t) \longrightarrow 0 \text{ uniformly,}$$

as $t \longrightarrow T$.

APPENDIX

A PROOF OF THE BONNESEN INEQUALITY

We present a proof of the Bonnesen inequality which applies to certain classes of locally convex, immersed curves as well as embedded ones. See [O2] and [S] for more on the Bonnesen inequality. Wallen has independently used this proof for a relative version of the Bonnesen inequality, see [W].

The proof relies on the formulae in terms of the support function for the geometric quantities associated to a curve and the Wirtinger and Poincaré inequalities. In the sequel p will denote the support function relative to a convenient choice of origin.

Proposition A.1: *For any constant r the following formula holds for a curve with total curvature $2\pi\nu$*:

$$\text{(A.1)} \qquad 1/2 \int_0^{2\pi\nu} [p_\theta^2 - (p-r)^2]d\theta = rL - A - \nu\pi r^2$$

Proof: Expand $(p - r)^2$ and apply formulae (3.3) and (3.4).

Lemma A.2: *The left hand side of (A.1) is positive if either*:

(a) $\int_0^{2\pi\nu} (p-r)d\theta = 0$ *i.e.*, $r = L/2\pi\nu$ *and*

$\int_0^{2\pi\nu} e^{i\ell\theta/\nu}p(\theta)d\theta = 0$ *for* $0 < \ell < \nu$.

or

53

(b) *the distance between successive zeros of* (p-r) *is always less than or equal to* π.

Proof: The Poincaré inequality states that if f is perpendicular to all the eigenfunctions of d^2/dx^2 with eigenvalues less than 1 then:

$$\int_0^{2\pi\nu} (f_\theta^2 - f^2)d\theta \geq 0.$$

Part (a) of the lemma follows immediately from this fact. Part (b) follows as well for if f is a function such that f(a) = f(b) = 0 and $|a{-}b| \leq \pi$ then

$$\int_a^b (f_\theta^2 - f^2)d\theta \geq 0.$$

This is, of course, the Wirtinger inequality. ■

Remark: For curves with a C^1 support function (A.1) is zero if and only if the support function is constant. The convex objects for which equality hold with $r = r_{in}$, such as the Minkowski sum of a non-trivial line segment and a circle, do not have C^1 support functions.

Proposition A.3: The Bonnesen Inequality
For a convex embedded curve

$$rL - A - \pi r^2 \geq 0 \ for \ r \in [r_{in}, r_{out}],$$

where r_{in} *is the radius of the largest inscribed circle and* r_{out} *the radius of the smallest circumscribed circle.*

Remark: In fact the Bonnesen inequality holds for all simple closed planar curves, however this proof requires that the curve be convex.

Proof: Since $rL - A - \pi r^2$ is a concave function it suffices to show that it is positive when $r = r_{in}$ and r_{out}. If we define the support function p relative to the center of the largest inscribed circle then the separation between two successive zeros of $p - r_{in}$ is at most π. Were this not the case then a simple variational argument shows that r_{in} is not the maximal inscribed radius. Similarly if we use the center of the smallest circumscribed circle to define p, then the zeros of $p - r_{out}$ are separated by at most π. In either case Lemma A.2 implies that $rL - A - \pi r^2 \geqslant 0$. ∎

This inequality has many equivalent forms, see [01-2]:

Corollary A.4: *For simple convex curves:*

(A.2)
$$L^2/A - 4\pi \geqslant \frac{\pi^2}{A} (r_{out} - r_{in})^2.$$

Proof: This follows by observing that the smaller root of $rL - A - \pi r^2 = 0$ is less than r_{in} while the larger root is larger than r_{out} and the applying the quadratic formula. ∎

A consequence used in §3 is

Corollary A.5 [G1]: *For a convex simple curve:*

$$LA \geqslant \pi \int_0^L p^2 ds.$$

Proof: For curves that are symmetric about the origin one has:

$$r_{in} \leqslant p \leqslant r_{out};$$

where p is defined relative to the center of symmetry. Hence

$$pL - A - \pi p^2 \geq 0.$$

If we integrate this relative to arclength we obtain the result for symmetric curves. In [G1] a "cut and paste" symmetrization argument is presented which reduces the theorem to the above special case; we refer the interested reader there for the details. ■

As a last corollary we state a Bonnesen type inequality for curves with a high order of symmetry. That is γ has total curvature $2\pi\nu$ and a q-fold symmetry such that $\nu/q < 1/2$. In §5 we showed that such a curve cannot pass through its center of symmetry. Define r_{in} and r_{out} as the radius of the largest inscribed circle about this point and the smallest circumscribed circle, respectively.

Let $p(\theta)$ be the support function relative to the center of symmetry. Evidently $r_{in} \leq p(\theta) \leq r_{out}$. On account of the symmetry

(A.3)
$$p(\theta + 2\pi\nu/q) = p(\theta).$$

By the definitions of r_{in} and r_{out} it follows that both $p(\theta) - r_{in}$ and $p(\theta) - r_{out}$ vanish somewhere. In virtue of (A.3) it follows that the zeros of these functions are separated by less than π. Thus Lemma A.2(b) implies:

Corollary A.6: *For a highly symmetric, locally convex curve:*

$$rL - A - \pi\nu r^2 \geq 0 \; for \; r \in [r_{in}, r_{out}].$$

In fact as $\nu/q < 1/2$ an even sharper inequality holds for if $r \in [r_{in}, r_{out}]$ the Poincaré inequality implies:

(A.4)
$$\int_0^{2\pi\nu} p_\theta^2 d\theta \geq \left[\frac{q}{2\nu}\right]^2 \int_0^{2\pi\nu} (p-r)^2 d\theta.$$

56

Therefore:

(A.5) $rL - A - \pi \nu r^2 \geqslant 1/2 [(q/2\nu)^2 - 1] \displaystyle\int_0^{2\pi\nu} (p-r)^2 d\theta \ \ for \ \ r \in [r_{in}, r_{out}].$

Bibliography

[Ab-La]   Abresch, U. and Langer, J., The normalized curve shortening flow and homothetic solutions, Jour. of Diff. Geo., 1986.

[Ch]   Chow, Bennet, Deforming convex hypersurfaces by the nth root of the Gauss curvature, J. of Diff. Geom., Vol. 22, No. 1 (1985), 117-138.

[D]   Deturck, D., Deforming metrics in the direction of their Ricci tensors, J. of Diff. Geom., Vol. 18 (1983), 157-162.

[Ep-We]   Epstein, C.L., Weinstein, M., A stable manifold theorem for the curve shortening equation, to appear CPAM.

[G1]   Gage, M., An isoperimetric inequality with applications to curve shortening, Duke Math. J., Vol 50, No. 4 (1983), 1225-29.

[G2]   ─────────, Curve shortening makes convex curves circular, Invent. Math. 76 (1984), 357-64.

[G3]   ─────────, An area preserving flow for plane curves, Contemp. Math., Vol 51, 1986.

[G-H]   ───────── and Hamilton, R., The shrinking of convex plane curves by the heat equation, Jour. of Diff. Geo., 1986.

[Hu]   Huisken, G., Flow by mean curvature of convex surfaces into spheres, Jour. of Diff. Geo., Vol 20, No. 1 (1984), 237-266.

[01]   Osserman, R., The isoperimetric inequality, Bulletin of
         the AMS, 1979.

[02]   —————————, Bonnesen style isoperimetric inequalities,
         Amer. Math. Monthly 86, No. 1 (1979).

[S]   Santalo, L., Geometric Probability and Integral Geometry,
         Addison Wesley, New York.

[W]   Wallen, L., All the way with the Wirtinger inequality,
         preprint.

LAX'S CONSTRUCTION OF PARAMETRICES OF FUNDAMENTAL SOLUTIONS OF STRONGLY HYPERBOLIC OPERATORS, ITS PREHISTORY AND POSTHISTORY

Lars Gårding

§1. Lax's parametrix construction

In the year of the Lord 1957 there appeared in volume 24 of Duke Journal a short paper by Peter Lax with the title 'Asymptotic solutions of oscillatory initial value problems'. It was one of the first signs that the aera of microlocal analysis was approaching. In this lecture I will try to look at Lax's paper in the context of what was before and what came after.

Its point of departure is the classical high frequency approximation to solutions of partial differential equations. Lax used them to show that hyperbolicity is a necessary property of partial differential equations with analytic coefficients in order to have good local Cauchy problems but he also constructed parametrices of fundamental solutions of variable coefficients strongly hyperbolic operators, in his case a first order operator

$$P(t,x,D_t,D) = ED_t + A(t,x,D) + A_0(t,x),$$

$$A(t,x,D) = \Sigma\ A_j(t,x)D_j\ .$$

Here E is the unit m×m matrix and the other coefficients are m×m matrices smoothly dependent on time t and on $x \in R^n$. D_t is $\partial/i\partial t$ and D is the imaginary gradient $\partial/i\partial x$. A fundamental solution of P is a distribution F(t,x) with the property that $PF(t,x) = \delta(t)\delta(x)$. One such can be obtained by putting F(t,x) = H(t)V(t,x) where H is the Heaviside function and V solves the Cauchy problem

Research supported in part by NSF Grant DMS-8120790.

$$PV(t,x) = 0, \quad t=0 \Rightarrow V = \delta(x).$$

By definition, a parametrix of F is a distribution G such that F–G is smooth, it can be obtained by putting G = HU where V–U is smooth.

In high frequency approximations one considers asymptotic series

(1) $$u = \exp i\lambda f(t,x) \; \Sigma \; g_k(t,x)\lambda^{-k}, \quad k=0,1,\ldots,$$

with real phase function f and possibly complex coefficients g_k, both smooth functions. One gets an asymptotic solution of Pu = 0 by requiring that

(2) $$\det(E\frac{\partial f}{\partial t} + A(t,x,\frac{\partial f}{\partial x}) = 0,$$

and that

(3) $$((E\frac{\partial f}{\partial t} + A(t,x,\frac{\partial f}{\partial x})g_0 = 0,$$

(4) $$((E\frac{\partial f}{\partial t} + A(t,x,\frac{\partial f}{\partial x})g_k + A(t,x,D_t,D)g_{k-1} = 0.$$

The strong hyperbolicity means that the equation (2) factors into m first order real non-linear Hamilton–Jacobi equations

(5) $$\frac{\partial f}{\partial t} + L_k(x,\frac{\partial f}{\partial x}) = 0.$$

For small t, they have unique solutions for which the values for t = 0 can be given arbitrarily. Corresponding to each solution $f = f_k$ there is a solution g_0 of (3), unique part from a scalar factor c(t,x). For the equation (4) for g_1 to be solvable, c has to satisfy a linear differential equation, the transport equation, and analogously for the other coefficients. This equation leaves the initial value c(0,x) free. The upshot of all this is that we can construct, for instance, an

61

asymptotic solution $U(t,x,\lambda)$ in the form of an m by m matrix which for t = 0 equals

$$U(0,x) = \exp ig(x)\lambda E,$$

with some arbitrary function $g(x)$ which is the common initial value of the solutions f_k of the equations (2).

So far the constructions were well known. The novelty of Lax's paper is that he replaces λ by n parameters

$$\xi = (\xi_1,...,\xi_n),$$

the phase function λf_k by function $f_k(x,\xi)$ homogeneous of degree one in ξ and the amplitudes $\lambda^{-k}g_k$ by functions $g_k(x,\xi)$ homogeneous of degree $-k$ in ξ. By the preceding calculations, this replaces the former U by an asymptotic series

(6) $$U(t,x,\xi) = U_0(t,x,\xi) + U_1(t,x,\xi) + ...$$

where U_k is homogeneous of degree $-k$ in ξ. We can also prescribe that

(7) $$U(0,x,\xi) = (2\pi)^{-n}E \exp ix \cdot \xi.$$

Integrating (7) with respect to large ξ produces a distribution which differs from $\delta(x)E$ by a smooth function. Hence the same integration of (6) should produce an asymptotic series representating a parametrix of the solution V above of PV = 0. In fact, since U_k is homogeneous of degree $-k$ in k, the terms of this series are arbitrarily smooth functions of t, x when k is sufficiently large. The singularities of the terms of the series can be determined very explicitly. Consider the instance a typical solumn of U,

$$H(t,x) = \int h(t,x,\xi) \ \exp if(t,x,\xi)d\xi,$$

with integration over, say, $|\xi| > 1$. If a point t,x is such that $f(t,x,\xi)$ can be taken as a coordinate on ξ space, integrations by parts and a partition of unity show that H is smooth close to that point. This shows that the integral is a smooth function outside those t,x for which the ξ gradient of f vanishes for some t. There are two ways of interpreting this condition. One is that t,x and 0 can be joined by the projection on t,x–space of some bicharacteristic

$$dt/ds = 1, \quad dx/ds = \partial f/\partial\xi, \quad d\xi/ds = -\partial f/\partial x, \quad \tau + f = 0.$$

The other is that the singularities of the integral H are concentrated to the envelopes of the family of surfaces $f(t,x,\xi) = 0$ for varying ξ. Since there is an involution $k \longrightarrow k'$ such that $f_{k'}(t,x,\xi) = f_k(t,x,-\xi)$, the m envelopes obtained in this way produce $[(m+1)/2]$ sheets in t,x–space issuing from t=0, x=0, which carry all the singularities of U(t,x) and hence also of the solution V(t,x) above of PV = 0. These are the main results.

For m=2 there is just one sheet precisely as for the wave equation. In his construction of a parametrix for the wave operator, Hadamard had used a geometrical method which amounts to guessing the equation of the sheet and the form of the solution and doing some verification afterwards. With more than one sheet carrying the singularities this method was hopeless. The new method was a triumph of the Fourier transform over the geometrical method.

Lax's article was preceded by a note in the Proceedings of the National Academy of Sciences by Courant and Lax where the same main results are announced but now using the Radon transform which is a decomposition of the δ-function into plane waves. This way comes a bit closer to the geometric method but lacks the freedom of the Fourier transform.

§2. The prehistory

Peter Lax's paper marks the second time in the history of hyperbolic equations when geometric methods proved insufficient and were replaced by a fearless use of the Fourier transform. The first instance is due to the Swedish mathematician Nils Zeilon and occurred in the wake of Sonya Kovalevskaya's abortive attempt to solve Cauchy's problem for the equations for the propagation of light in doubly refracting crystals. This problem has an interesting history.

The theory of caustics was developed very successfully in the beginning of the nineteenth century by Hamilton under the name of geometrical optics. At about the same time, Poisson found that the function

$$ u(t,x) = (4\pi)^{-1} \int f(y)\delta(t-|x-y|) \frac{dy}{|x-y|} $$

where dim x,y = 3 solves the following Cauchy problem for the wave equation,

$$ u_{tt} - \Delta u = 0; \quad u = 0, \quad u_t = f(x) \text{ when } t = 0. $$

Its fit with geometric optics was perfect, the support of the solution at time t is precisely the envelope of the sphere with radii t centered in the support of f. The spherical waves $F(t-|x|)/|x|$ with F arbitrary were also known as solutions of the wave equation.

These results led to the idea that geometrical optics represented all of wave propagation. At the same time there were efforts to analyze light as kind of vibration in some more or less mysterious substance. One test for this theory was the propagation of light in doubly refracting crystals. It had been known for a long time that certain crystals break an incoming beam of light into two beams whose directions vary with the direction of the incoming beam relatively to the crystal. According to Huygens's theory this means that light in the crystal has two propagation velocities, both direction dependent. When these velocities were measured for Iceland spar, it turned out that they were unequal except for two directions, the two optical axes of the crystals. The explicit form of the corresponding velocity surface, i.e. the velocities as functions of directions, was a problem for a long time until the French physicist Fresnel found its analytic form, a fourth degree surface depending on three parameters. By the principles of geometrical optics, the corresponding wave surface, i.e. the wave front formed at a fixed time by light from a point source in the crystal, is the envelope of the velocity surface. By a freak of nature, the wave surface is obtained from the velocity surface just by inverting the parameters. Fresnel only guessed the form of the wave surface, others carried out the computations, among them Hamilton. Like the velocity surface, the wave surface is symmetric around the origin and has four double points. The way to think of it is to imagine two balloons, one inside the other and glued together at four points. When the inner balloon and the space between them has been blown up, they assume the form of the wave surface.

In his study of Fresnel's wave surface, Hamilton predicted that a beam of light entering the crystal along an optical axis ought to be broken into a conical beam. In fact, the tangents to the velocity surface at a double point form a plane tangent to the wave surface, part of which is a lid covering the inlet to a double point of the wave surface. At the boundary of the lid, the corresponding tangent planes through a double point of the velocity surfaces are strictly tangent to it (and do not only pass through the double point). Hamilton observed this phenomenon and predicted that a beam of light entering a crystal from the ouside in the direction of an optical axis should be broken into a cone of light beams. A short time afterwards, this prediction was verified by experiment. As it turned out later, the complete wave surface is obtained by taking the convex hull of the outer sheet of the Fresnel surface. This provides it with four lids as described above. In the development that followed, conical refractions was forgotten for a long time. More pressing business waited, the theory of light waves rather than light rays.

The model first chosen for light waves was based on analogies with elasticity. The French mathematical physicist Lamé devised a

65

3×3 hyperbolic system of second order equations which ought to describe the intensity of light and its propagation in doubly refracting crystals. He was successful. His system is precisely what one gets out of Maxwell's equations by eliminating the magnetic field. Lamé also found a solution of his system analogous to the spherical waves of the wave equation. It has the right connection with the wave front surface but is infinite on the optical axes.

To solve Cauchy's problem for Lamé's system was a great challenge. The first to try it was Sonya Kovalevskaya. Her point of departure was a manuscript by Weierstrass where he solved Cauchy's problem for a system analogous to that of Lamé but associated with the product of two wave operators with different speeds of light. She used the tools at her disposal, namely Lamé's solution, Weierstrass's manuscript and the idea that light should be concentrated between the two sheets of the wave surface. This was in fact what Weierstrass had found. Although she voiced the usual complaints about Lamé's solution, she proceeded to use it and the the parametrization of the wave surface by elliptic functions found earlier by H. Weber.

Some years after the publication of Kovalevskaya's article and after her death, Vito Volterra discovered that her solution was identically zero, a clear contradiction with the Cauchy–Kovalevskaya theorem. Trying to remedy the situation, he constructed solutions of Lamé's system outside the plane of the optical axes depending on a number of arbitrary functions, but he could not solve Cauchy's problem. He announced it as an important mystery in a series of lectures that he gave in Stockholm in 1906. One of his listeners was a young Swedish mathematician, Nils Zeilon. His admired teacher Fredholm had constructed fundamental solutions of elliptic constant coefficient partial differential operators in three variables using Abelian integrals. Zeilon found another point of departure, namely the Fourier integral. He refers to Cauchy for the formula

$$E(x) = (2\pi)^{-n} \int \exp ix \cdot \xi \ d\xi / P(\xi)$$

for a fundamental solution E of a partial differential operator P(D). In 1911 he published a paper where he tried to give a sense to the integral when P is homogeneous and n=3. He arrived at Fredholm's formulas extended to non-elliptic operators. His computations are questionable, but the results are right. Zeilon, who also took an interest in hydrodynamics, was more of a mathematical physicist than a mathematician. His success with the Fourier transform and fundamental solutions gave him tools to attack the problem of double refraction. He published an article in 1919 about fundamental solutions for n=4 followed by another one in 1921 where he solved the problem completely. He found the precise support of the solution of Lamé's system with given Cauchy data and analyzed its singularities completely. At the very end he remarks about the lids of the Fresnel wave surface that they prevent light from escaping as it were. The precise interpretation is of course that the velocity of light in the conical beam is obtained by linear interpolation. Zeilon's remark shows that he was not familiar with conical refraction.

Only the first of Zeilon's articles reached Jahrbuch über die Fortschritte der Mathematik. Therefore he is not mentioned by his successors Herglotz and Petrovsky. The connection between the support of the fundamental solution of Lamé's system (or rather that of a closely connected fourth order scalar equation) and conical refraction remained obscure until 1961 when a paper by Ludwig clarified the situation.

§3. The posthistory

Let me now look at what happened after Lax's paper. There are two points to make, the first one is about the interpretation of what Courant and Lax called the generalized Huygens's principle. Experiences with hyperbolic equations in two variables led to the belief early on that the singularities of homogeneous hyperbolic equations should propagate along bicharacteristics. Courant and Lax called this the generalized Huygens's principle. Lax's construction proves the principle for his hyperbolic systems. But this has not been the end.

In microlocal analysis, singularities of a distribution u at a point are measured by the size in various directions of the Fourier transform of u, localized at the point. More precisely, one defines a function $r(x,\xi) = r_u(x,\xi)$ which may be called the regularity function of u with the following property: it is the greatest lower bound of numbers m for which the Fourier transform of hu with h a smooth function with support close to x, $h(x) \neq 0$ is square integrable far away in a narrow cone around ξ when multiplied by $|\eta|^m$. Loosely speaking, the Fourier transform of u at x has the size $|\eta|^{-m}$ conically close to ξ when m is close to $r(x,\xi)$. The generalized Huygens principle now takes the form of Hörmander's basic propagation of singularities theorem (1970). It runs as follows. Let a distribution u solve a linear equation

$$P(x,D)u = 0,$$

and suppose that the characteristic polynomial $Q(x,\xi)$ of the principal part of P is real of principal type, i.e. it is real for real ξ and its gradient does not vanish outside the origin. Then the regularity function of P is infinite when $Q(x,\xi)$ is not zero and constant along the bicharacteristics of P, in this case simply the solutions of the Hamiltonian system

$$dx/dt = \partial Q/\partial \xi, \quad d\xi/dt = -\partial Q/\partial x, \quad Q(x,\xi) = 0.$$

This direct coupling between singularities and bicharacteristics is one of the great achievements of microlocal analysis. The bicharacteristics appear as trajectories of classical particles and microlocal analysis as an operator calculus on the phase space of classical mechanics, the product of physical space with coordinate x and momentum space with coordinate ξ.

The second point is about a defect of Lax's construction which is shared by Hadamard's construction of a parametrix for wave equations with variable coefficients. Both of them are just local.

The parametrices are constructed only close to the pole. The reason in the general case is that the Hamilton–Jacobi equation does not have global solutions except in special cases.

Soon after Lax's paper Ludwig indicated that his construction could be made global if cut up into steps, but the construction was very implicit and uses oscillatory integrals with more and more integration variables as time increases. Hörmander and Duistermaat met the challenge in their joint paper Fourier Integral Operators II. Away from the origin, the problem is reduced to extending everyone of the m terms of the parametrix with different phase functions individually along the bicharacteristics. Hörmander and Duistermaat could do this by partitions of unity in phase space and by using their realizations of canonical transformations as Fourier integral operators. It is, however, possible to avoid the Fourier integral operators and to extend Lax's construction to arbitrarily large times just by suitable changes of variables. For a single pseudodifferential equation of order 1, this has been done in the lecture I has given recently in Tientjin at the Institute of Mathematics of Nankai University. The construction is very simple and can be sketched in a few lines.

Consider a first order classical pseudodifferential operator with smooth coefficients

$$A = D_t + P(t,x,D), \qquad D = D_x,$$

defined on the product of the real line and R^n. Suppose that A is hyperbolic in the sense that its principal symbol

$$\tau + q(t,x,\xi)$$

is real for real argument. Consider the corresponding bicharacteristics $t \longrightarrow x(t)$, $\xi(t)$ defined by

$$dx/dt = \partial q/\partial \xi, \qquad d\xi/dt = -\partial q/\partial x, \qquad \tau = -q,$$

for $t>0$, issuing from y, η for $t=0$ and $\eta \neq 0$. If we assume that the

69

bicharacteristic flow maps bounded sets of $T^*(R^{2n})$ into bounded sets, there is a distribution $E(t,x)$ which is a fundamental solution modulo smooth functions,

$$AE(t,x)-\delta(t)\delta(x) \text{ smooth for } t \geq 0, \quad E=0 \text{ for } t<0.$$

Precisely as in Lax's construction, we can find a phase function $h(t,x,\eta)$, solution of the Hamilton–Jacobi equation,

$$\partial h/\partial t + q(t,x,\frac{\partial h}{\partial x}) = 0, \quad h = y \cdot \eta \text{ when } t = 0,$$

and an amplitude $a(t,x,\eta)$ of degree 0, equal to $1/(2\pi)^n$ for $t=0$, such that the oscillatory integral

$$F(t,x) = \int a(t,x,\eta)\exp ih(t,x,\eta) \ d\eta$$

differs from $E(t,x)$ by a smooth function when t is sufficiently small. Since h is also defined by the property that

$$y \cdot \eta = h(t,x(t),\eta),$$

the construction works as long as the map

$$(8) \qquad\qquad y,\eta \longrightarrow x(t),\eta$$

is invertible. Since the map $y,\eta \longrightarrow x(t),\xi(t)$ is also invertible, we can introduce the variables $\xi = \xi(t)$ in the integral F getting

$$(9) \qquad\qquad F(t,x) = \int b(t,x,\xi)\exp if(t,x,\xi) \ d\xi,$$

still when $t>0$ is small. Here

$$(10) \qquad\qquad y \cdot \eta = h(t,x(t),\eta) = f(t,x(t),\xi(t)).$$

70

Let us also note that

$$\eta.dy = \xi \cdot dx \quad \text{when} \quad dt = 0$$

and that $\xi \cdot dx = 0$ defines a homogeneous Lagrangian manifold L(t).
In particular, $L(0) = 0 \times R^n \backslash 0$ is the wave front set of E(0,x). Hence
L(t) is the wave front set of $x \longrightarrow E(t,x)$ and the wave front set of E is
the union of these manifolds augmented with the pairs t, τ. In order
to make the parametrix global, we shall make a partition of unity of a
conical neighborhood of L(0) in the cotangent bundle of R^n. Via the
bicharacteristic map we then get partitions of unity of corresponding
neighborhoods of the L(t). Let N(0) be a conical neighborhood of an
element $0, \eta$ of N(0), let B: $t \longrightarrow x'(t), \xi'(t)$ be the corresponding
bicharacteristic and N(t) the image of N(0). It is clear that the map
(8) related to N(0) and N(t) is invertible for larger values of t when
N(0) decreases but there is nothing that says that all values of t are
attainable in this way. On the other hand, it is also possible to try
to prolong the parametrix in the form (9), starting at time t and using
the procedure that carried us from 0 to t. This seems advantageous
since (10) shows that the function $f(t,x,\xi)$ is defined for all t. The
only hitch in this program turns out to occur at a time t when the
variable $\xi(t)$ is no longer a parameter on L(t) as η was at the
outset). But this defect is remedied by a change of the variables x
close to x'(t) and a concomitant change of the ξ so that $\xi \cdot dx$
remains invariant. This change of variables is of course only local but
can be made to cover N(t) after a suitable reduction of N(0). The
stage is now set for a prolongation of the integral for larger values
of t. A routine covering argument completes the proof. The result is
that for any given t, there is a parametrix in the form of a locally
finite sum of oscillatory integrals of the form (9). This result is the
same as that found by Hörmander and Duistermaat but the proof is
simpler. It shows that Peter Lax's construction of parametrices for
fundamental solutions of hyperbolic operators is inherently global.

REFERENCES

Courant, R., and Lax, P.D., The propagation of discontinuities in wave motion, Proc. Nat. Acad. Sc. 42 (1956), 872-876.

Duistermaat, J.J. and Hörmander, L., Fourier integral operators II, Acta Math. 128 (1972), 183-269.

Fresnel, A.J., Oeuvres complètes (1866-1870), Paris, Imprimerie Impériale.

Gårding, L., Singularities in linear wave propagation, Lectures 1986 at the Nankai Institute of Mathematics, Tientsin.

Hadamard, J., Le problème de Cauchy, Hermann, Paris, 1932.

Hamilton, W.R., The Mathematical Papers of Sir William Rowan Hamilton I, II, III, eds. Conway, Synge, Cambridge Univ. Press (1931), I, 280-293.

Herglotz, G., Über die Integration linearer partieller Differentialgleichungen mit konstanten Koeffizienten I-III, Ber. Sächs Akad. Wiss. 78 (1926), 93-126, 287-318, 80, (1928), 69-114.

Hörmander, L., Linear differential operators, Actes Congr. Int. Math. Nice 1970, 1, 121-133 and On the existence and regularity of solutions of linear pseudodifferential equations, Ens. Math. 17, (1971), 99-163.

-- Fourier integral operators I, Acta Math. 127 (1971), 79-183.

Kovalevskaya, S.V., Über die Brechung des Lichtes in crystallinschen Mitteln. Acta Math. 6 (1885), 249-304.

Lamé, G., Lecons sur le théorie de l'Elasticité des Corps Solides, Deuxième ed. (1866), Paris, Gauthier-Villars.

Lax, P.D., Asymptotic solutions of oscillatory initial value problem, Duke J. 24 (1957), 627-646.

Ludwig, D., Exact and asymptotic solutions of the Cauchy problem, Comm. Pure Appl. Math. 13 (1960), 473-508.

Ludwig, D., Conical Refraction in Crystal Optics and Hydromagnetics, Comm. Pure and Appl. Math. XIV (1961), 113-124.

Petrovsky, I.G., On the diffusion of waves and the lacunas for hyperbolic equations, Mat. Sb. 17(59), (1945), 289-370.

Weber, H., Über die Kummersche Fläche vierter ordnung... Journal F. Math. 84 (1978), 332-354.

Volterra, V., Sur les vibrations lumineuses dans les milieux birefringents, Acta Math. 16 (1892), 154-215.

-- Lecons sur l'intégration des équations différentielles aux derivées partielles professées à Stockholm (1906), Upsal 1906, Almqvist O. Wicksell.

Zeilon, N., Das Fundamentalintegral der Allgemeinen Partiellen Differentialgleichung mit Konstanen Koffizienten, Arkiv. F. Math. Fys. O. Astr. 6 (1911), No.38.

Zeilon, N., Sur les équations aux derivées partielles à quatre dimensions et le problème optique des milieux birefringents I, II, Nova Acta Reg. Soc. Sc. Upsaliensis, Ser.IV, 5, No.3, (1919), 1-57 and No.4 (1921), 1-130.

THE BUKIET-JONES THEORY OF WEAK DETONATION WAVES IN CURVILINEAR GEOMETRY

James Glimm[1,2,3]

Courant Institute, New York University
New York, N.Y. 10012
and
Los Alamos National Laboratory
Los Alamos, N.M. 87545

Abstract

Front tracking is a numerical method which uses exact mathematical knowledge of idealized solution singularities within the computational process. This fact is typical of a broadly based scientific motivation for the study of nonlinear waves, wave interactions and Riemann problems in one and higher dimensions. Such a study is pertinent for a variety of physical systems, including oil reservoirs, real materials with realistic equation of state and constitute relations, chemically reactive fluids and plasmas.

In this note, we report on a recent results of J. Jones in formulating and analyzing model equations to understand qualitatively and quantitatively (to first order) the effect of curvature on an expanding detonation wave front and the validation by numerical methods of the asymptotic correctness of this theory by B. Bukiet. These results are a necessary input to the front tracking method.

1.   Supported in part by the National Science Foundation, grant DMS-831229.
2.   Supported in part by the Applied Mathematical Sciences subprogram of the Office of Energy Research, U.S. Department of Energy, under contract DE-AC02-76ER03077.
3.   Supported in part by the Army Reserach Office, grant DAAG29-85-K-0188.
      Research supported in part by NSF Grant DMS-8120790.

1. Introduction

The characteristic difficulty in the modeling of detonation waves is the strong degree of couping which is exhibited between the fluid and chemical modes, so that the problem cannot be effectively reduced to an understanding of the constituent subsystems. The central difficulty in the computation of detonation waves is the stiffness, steep gradients, or narrow combustion zones, with large changes in essential quantities occurring over a small fraction of a typical computational domain. From the point of view of chemistry, the reaction rates, the intermediate reaction products and their equation of state are not understood on a fundamental level. There are furthermore instabilities in detonation waves; the basic mode arises from a tendency of the chemistry to oscillate between lagging behind and catching up to the fluid modes. This basic unstable mode occurs in one space dimension as a time dependent oscillation (the galloping detonation), in two space dimensions as crinkled wave fronts, Mach stems and transverse or surface waves and in three dimensions as heterogeneities, voids and hot spots. A good introduction to this range of issues can be found in [7,8,11].

The problem which Bukiet and Jones addressed was the effect of wave front curvature on detonation speed. An expanding wave front or more generally a diverging flow field or rarefaction wave generally produces a lowering of pressure, temperature and density. This is a purely fluid phenomena, and follows from the basic conservation laws such as conservation of energy. For the case of a chemically reacting fluid, the result of these changes is to decrease the chemical reaction rates. This is an instance of the coupling of the chemical to the fluid modes. Because of the converse coupling of the fluid to the chemical modes, the result of the decreased chemical activity is a decrease in detonation wave speed.

One dimensional (planar) waves can be reduced to a stationary frame, in which case the governing equations reduce to ordinary differential equations. For detonation waves the resulting equations are known as the ZND theory. This theory predicts a one parameter family of solutions, with the relevant parameter determined by a

75

downstream boundary condition, such as pressure. This family terminates at a point (called the CJ point) where the downstream flow field becomes sonic at the end of the reaction zone. Excluding the (CJ) endpoint, the detonation waves in this family are called strong, or supported because the detonation velocity is determined by (supported by) a downstream boundary condition. The CJ detonation at the extreme end of this family is called unsupported; and the flow field beyond the sonic point has no influence on the detonation and thus in this case the detonation velocity is independent of downstream boundary conditions.

On the basis of conservation law jump relations (Rankine-Hugoniot relations), possible detonation wave and flame velocities and jump transitions can be analyzed [6]. In this analysis, there are two branches, one for flames (deflagrations) and one for detonations. The branch associated with detonation waves can be continued past the CJ point to what are known as weak detonation waves. In the context of planar waves and exothermic reactions, these weak waves can be regarded as unstable and hence unphysical [6]. However detonation waves which are at least qualitatively similar to the weak waves of this branch do arise and Fickett and Davis [7] explain the mathematical mechanisms for this to occur. As mentioned above, endothermic effects, such as an expanding wave front, lead to slower chemistry and a decrease in velocity below the CJ point and the desired detonation trajectory is then one of the separatrices associated with this saddle, at least in simplified models [7].

In order to address the basic problem of computational methods for combustion as mentioned above (stiffness or incomparable length scales), it was necessary to turn the above ideas tested in model equations into a theory which was (a) directly connected with the equations of chemically reactive flow and (b) quantitative as a correction to the planar ZND theory. These two steps have been taken by Bukiet [3,4,5] and Jones [9,10] and are the subject of this article. On the basis of their accomplishment, interface methods in general and front tracking in particular can be used in the numerical computation of detonation waves. The importance of this problem has been underscored by C. Mader [11].

The outlines of the work is as follows: Jones derived a set of equations which are equivalent to first order in the curvature to the Euler equations of a compressible, reactive fluid in the large time, small curvature limit. These equations are 2×2 systems of autonomous ordinary differential equations. The derivation is formal in that a solution with the required regularity properties is postulated. Deep questions concerning the asymptotic behavior of partial differential equations which would be part of a fully rigorous derivation of the model equations from the full Euler equations are not resolved. Bukiet established the validity of these equations by a comparison of their solution with a numerical solution of the full Euler equations using the simplifications of radial symmetry. Jones has analyzed the mathematical structure of the autonomous ordinary differential equations. Rigorous proofs are somewhat rare in the area of asymptotic expansions, and to illustrate the importance of the numerical validation of the expansion, we note that Jones' equations inspired discussions as to whether the expansion should have terms proportional to the log or a fractional power of the curvature, and thus whether the true correction to the detonation velocity was linear in the curvature to leading order. In view of the difficulty of a rigorous mathematical resolution of these questions, the numerical validation was judged to be essential. Once completed, the numerical tests demonstrated that Jones' equations [9,10] were correct.

It is a pleasure to thank W. Fickett for helpful comments on an unpublished manuscript of J. Jones containing Jones' equations and for his transmission of this manuscript in the summer of 1985 to J. Bdzil for further comments. We also thank D.H. Sharp, R. Menikoff and C. Mader for helpful comments.

In a series of papers, Bdzil and co-authors have considered the related question of the cylindrical geometry. The studies of the steady detonation velocity are distinct from the problem considered here. In fact this problem is strictly time independent but has the extra compliation that the shape of the detonation front is not known a priori. Very recently, Bdzil and Stewart [1,2] have extended this steady analysis to include transients, but their work is still distinct from the considerations of Bukiet and Jones. Not only do they exclude Arrhenius kinetics, and require that the distance from the initiating shock to the sonic point be bounded uniformly as a function of the radius of curvature, but they take the sonic point to be the end of the reaction zone. Their methods require an integration from the sonic locus to the shock front to determine relations which yield the partial differential equation for the shock shape. This is similar in spirit to the use of a shooting method to proceed from the critical sonic point back to the beginning of the reaction zone, in what Fickett refers to as an eigenvalue problem for the detonation velocity. In any case Bdzil and Stewart confirm in their context the earlier discovery of Jones that the leading corrections to detonation velocity are linear in the curvature.

2. The Equations of Reactive Fluid Flow

We consider the Euler equations of reactive fluid flow in a radial geometry. These equations differ from the ordinary Euler equations of compressible gas dynamics by the presence of source terms and by the presence of a new equation. The new equation describes the reaction progress and corresponds to a simplified chemistry in which burning is accomplished by the release of heat in proportion to the amount of burned gas, and the reaction progress is this (mass) fraction of burned gas. The source terms arise from the combustion and also from the radial geometry. The equations are

$$\mathbf{w}_t + f(\mathbf{w})_r = C - G,$$

where

$$\mathbf{w} = \begin{bmatrix} \rho \\ m \\ e \\ \lambda \end{bmatrix} \qquad f(\mathbf{w}) = \begin{bmatrix} m \\ \dfrac{m^2}{\rho} + P \\ \dfrac{m}{\rho}(e+P) \\ m\dfrac{\lambda}{\rho} \end{bmatrix}$$

78

$$C = \begin{bmatrix} 0 \\ 0 \\ 0 \\ R(\lambda, T) \end{bmatrix} \quad G = \begin{bmatrix} \frac{m}{r} \\ \frac{m^2}{\rho r} \\ \frac{m}{\rho r}(e+P) \\ 0 \end{bmatrix}$$

Here ρ is the density, m is the momentum density, P is the pressure and λ is the mass fraction of burned gas. We assume in the following a polytropic equation of state and Arrhenius kinetics, in which case the energy per unit volume is

$$e = \frac{P}{\gamma - 1} + (1-\lambda)\rho q + \frac{\rho u^2}{2}$$

The heat released during combustion is q and with $T = \frac{P}{\rho}$ as temperature, the reaction rate is

$$R(\lambda, T) = k(1-\lambda)\exp\left[-\frac{E}{T}\right]$$

for $T \geqslant T_c$, where k is the rate multiplier, E is the activation energy and T_c is a critical temperature below which the reaction rate is taken to be identically zero.

3. Jones' Equations

The zero order equations give the ZND theory of a steady planar detonation wave. In first order, time enters only as a parameter, in the determination of the radius which enters into the geometrical source terms. Because of this, the time derivatives drop out of the first order equations, and a quasi steady 4×4 autonomous system of ordinary differential equations results. Two exact integrals of motion can be found which allows a reduction to a 2×2 system, to be analyzed by phase plane methods. In this system the dependent variables are λ (reaction progress) and v (steady frame velocity). A

79

change of independent variables is also performed to eliminate a singularity at the sonic point and to simplify the form of the equations. At this point the system has a unique critical point and the solution which corresponds to an unsupported detonation is the trajectory starting on the $\lambda = 0$ axis and running into the critical point. In detail, the equations are [9,10]

$$v_y = q(\gamma-1)k(1-\lambda)v - \frac{(\dot{z}-v)v}{z} c^2 \exp\left(\frac{E\gamma}{c^2}\right)$$

$$\lambda_y = k(1-\lambda)(c^2-v^2)$$

where

$$c^2 = c_a^2 + \frac{\gamma-1}{2}(\dot{z}^2-v^2) + q(\gamma-1)\lambda.$$

4. Bukiet's Validation of Jones' Equations

The solution of the separatrix through the critical point in Jones' equation was found by a shooting method. The comparison solution for the cylindrically symmetric reactive Euler equations was solved by the random choice method with the source terms added by operator splitting. Because this problem is effectively in one space dimension, it was possible to obtain full numerical convergence and to include several hundred mesh points within the reaction zone. Thus one can be confident of the accuracy of the comparison solution, which we now take as exact. For a more detailed discussion, we refer to [4].

To show the comparison which gives the correstness of Jones' equations, we present two figures from [4]. In Figure 1, the time evolution of the pressure behind the shock at the edge of the detonation wave is plotted, for three cases. One is a planar theory, as a solution to the ZND equation. This plot is distinct from the others, as it should be, but suggests agreement at infinite radius, as also should be the case. The other two curves are the Jones' theory and the comparison solution. They disagree for initial time due to differences in the initialization and to nonleading order effects in the

80

two computations, but show agreement at large times. In Figure 2, the pressure is plotted as a function of the curvature, and the agreement for small curvature is even clearer now. It is interesting to note that the pure leading order contribution, taken from Jones' equations, shows better agreement for moderate radii than does the full theory. There is no theoretical reason for or against this occurrence, but it does illustrate the importance of the higher order terms in the expansions and equations.

5. Jones' Analysis of his Equations

For $z = \infty$, the equations of Section 3 can be integrated exactly. There is a line of critical points at $\lambda = 1$, and the intersection of this line with the sonic locus is the primary bifurcation point of the system. This intersection has a bifurcation to a unique critical point, which is a saddle point, for finite z.

Theorem. Let $z < \infty$, but sufficiently large and assume that $T \geqslant T_c$ throughout the region behind the initial shock. The system of Section 3 has a unique critical point in the physical domain $0 \leqslant \lambda \leqslant 1$. This critical point is a saddle point and its coordinates in the v, λ plane are analytic functions of $\frac{1}{z}$.

81

Pressure

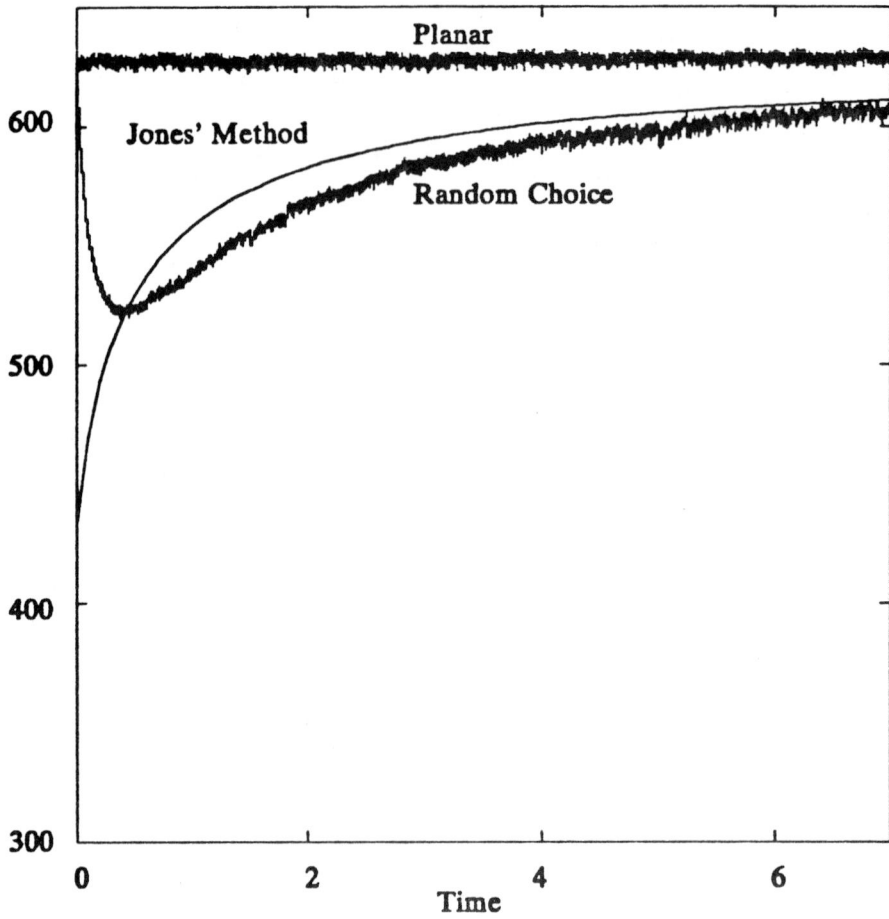

Fig. 1. Effect of Curvature on Pressure behind the Initiating Shock. *A plot of pressure vs. time for planar and cylindrical computations using the random choice method and the solution to Jones' equations where the radius of curvature is assumed to be the same as for the cylindrical run by random choice. The error bars show the range of values over each 100 time steps for the random choice calculations. The ahead state has $p = 300$, $u = 0$, $\rho = 1.4$. The reaction zone has length 1, $q = 300$, $T_c = 215$, $\gamma = 1.1$, $E = 100$ and grid spacing 0.01.*

Pressure

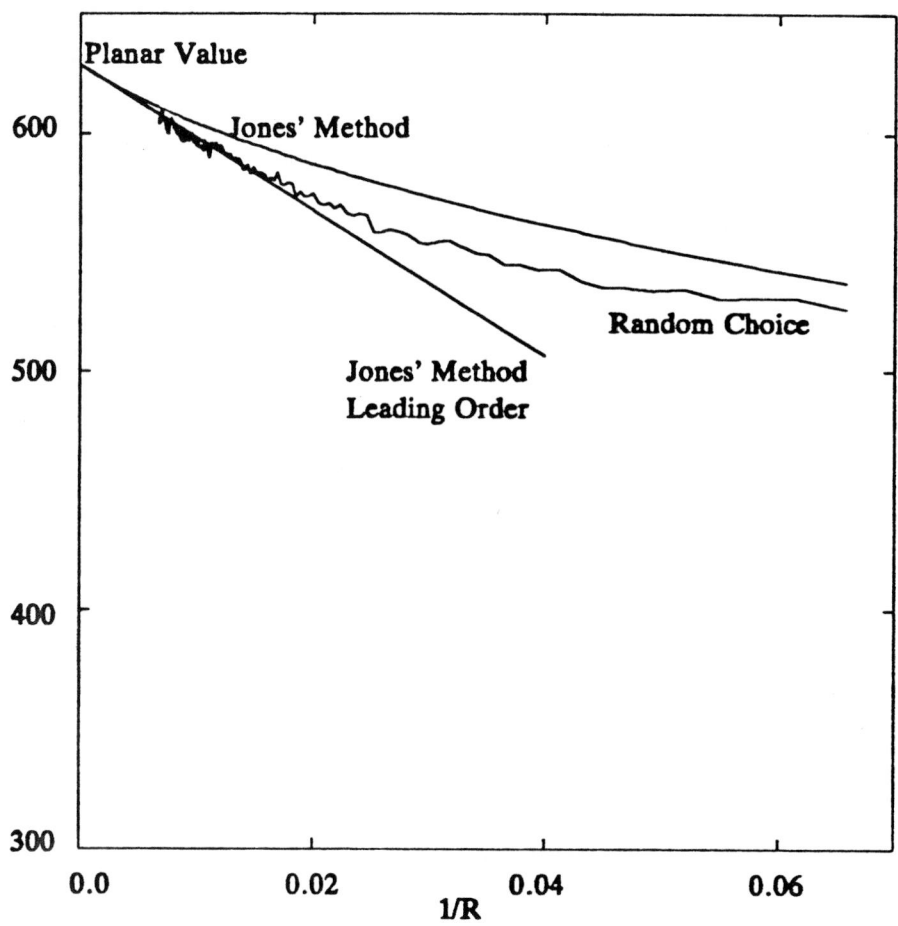

Fig. 2. First Order Corrections to Pressure. *Pressure behind the initiating shock wave is plotted against inverse radius for the cylindrical computations of* Fig. 1. *Also shown is the leading order correction predicted by solving Jones' equations for very large radii.*

REFERENCES

1. J.B. Bdzil and S. Stewart, Time-Dependent Two-Dimensional Detonation: The Interaction of Edge Rarefactions with Finite Length Reaction Zones, U. Ill., Preprint.

2. J.B. Bdzil and S. Stewart, The Shock Dynamics of Stable Multidimensional Detonation, U. Ill., Preprint.

3. B. Bukiet, Application of Front Tracking to Two Dimensional Curved Detonation Fronts, Submitted to SIAM J. Sci. Stat. Comp.

4. B. Bukiet, The Effect of Curvature on Detonation Speed, NYU, Preprint.

5. B. Buket, A Study of Some Numerical Methods for Two Dimensional Curved Detonation Problems, NYU Thesis, 1986.

6. R. Courant and K.O. Friedrichs, Supersonic Flow and Shock Waves, Springer-Verlag, New York, 1976.

7. W. Fickett and W.C. Davis, Detonation, University of California Press, Berkeley, 1979.

8. W. Fickett, Introduction to Detonation Theory, University of California Press, Berkeley, 1985.

9. J. Jones, Asymptotic Analysis of an Expanding Detonation, To Appear.

10. J. Jones, The Competition Between Curvilinear Geometry and Chemistry in Reactive Fluid Flow, Proceedings of the VIII International Congress on Mathematical Physics, July 16–25, 1986, Marseille.

11. C.L. Mader, Numerical Modeling of Detonation, University of California Press, Berkeley, 1979.

THREE-DIMENSIONAL FLUID DYNAMICS IN A
TWO-DIMENSIONAL AMOUNT OF CENTRAL MEMORY

Steven Greenberg, David M. McQueen, and Charles S. Peskin
Courant Institute of Mathematical Sciences
New York University
251 Mercer Street, New York, NY 10012

Dedication

To Peter Lax, with thanks for the Lax Report, which has led to expanded availability of supercomputers for scientific research, and with further thanks for letting CSP practice Mathematics without a license.

ABSTRACT

The subject of this paper is an out-of-core implementation of Chorin's projection method for the three-dimensional, time-dependent, incompressible Navier-Stokes equations in a periodic box. The implementation is particularly suitable for an architecture such as that of the Cray X-MP/SSD, in which a moderate-sized central memory is coupled to a much larger sequential-access memory. The central-memory requirements are only proportional to N^2, not N^3. Asynchronous i/o techniques are used to overlap i/o with computation and also i/o with itself. All inner loops are successfully vectorized by the Cray Fortran compiler. On the Cray X-MP/SSD, in the case $N = 64$, the implementation achieves a balanced distribution of cost between i/o and computation. At $N = 128$, the i/o efficiency is further increased, and the effective megaflop rate (megaflops/wall time) on a single processor with a dedicated machine is 77 megaflops/sec.

Research supported in part by NSF Grant DMS-8120790.

Introduction

The memory requirements of three-dimensional computational fluid dynamics are very severe. A computational lattice with 64 points in each direction contains a total of $64^3 = 2.5 \times 10^5$ lattice points. At each lattice point there are four fluid variables: the three components of the velocity and the pressure. Thus the minimal memory requirements for such a computation are about 10^6 words. In reality, the situation is much worse than this because several auxiliary arrays are needed during the computation. Moreover, in three-dimensional computation, lattice refinement by a factor of 2 causes an 8-fold increase in the required storage, and such lattice refinement is needed to provide empirical evidence of convergence.

In the design of advanced computers, there appears to be a tradeoff between central memory size and speed of computation. In the Cray X-MP and its descendents, for example, high speed is achieved by restricting the size of the central memory to a few million words. For users that need more than this, a Solid-state Storage Device (SSD) is provided. The SSD appears to the user as a very fast disk: Like a disk, it has the property that its use is most efficient when large amounts of data are transferred at one time. Moreover, such transfer can be carried out in *asynchronous* mode. That is, once the transfer is initiated, the central processor(s) can proceed with the computation while the transfer is carried out.

The Cray 2, on the other hand, is dedicated to the proposition that all data are created equal and should reside in a common, random-access central memory. Thus, a central memory of 256 million words is provided. Even in the Cray 2, however, there is a small, faster local memory associated with each of the 4 processors. This feature of the design (which is contrary to the philosophy expressed above) shows that it is not practical to provide equally fast access to all of the data that are stored in the machine. Inevitably there is a hierarchy in which the fastest access is to the registers of the central processor(s) and the slowest is to external devices such as disks. In this hierarchy, the large central memory of the Cray 2 actually holds an intermediate position.

The algorithm described in this paper is designed for a machine

86

with a small, fast, random-access "central" memory and a large, slower, sequential access "disc" memory. We assume also that asynchronous information transfer (asynchronous i/o) is possible between the central memory and the disc. If the three-dimensional computational lattice has N points in each direction, then the central-memory requirements of the algorithm are only proportional to N^2, while the external-memory requirements are proportional to N^3. Thus the central-memory requirements are only two-dimensional; this explains the title of the paper.

Whether the algorithm of this paper is appropriate for any particular machine depends, of course, on the size of the problem. For problems of the size reported here, the algorithm is appropriate for a Cray 1 or a Cray X-MP but not for a Cray 2. It should be mentioned, though, that the $code$ is still transportable to a machine such as the Cray 2 where the entire problem fits into central memory. All that is necessary is to declare that the "disc" files are to be treated as internal files. Moreover, depending on the charging algorithm, it may turn out that it is less expensive to use only a small amount of central memory even when a large central memory is available. Suppose, for example, that the cost of a computation is proportional to the product of the central-processor time and the amount of central memory occupied by the job. The method described in this paper dramatically reduces the amount of central memory (from $O(N^3)$ to $O(N^2)$) without increasing the central-processor time. Thus, a dramatic reduction in cost can be expected (with this particular charging algorithm). If there is also a charge for input/output, this will partly offset the savings. In practice, though, i/o is relatively cheap.

Ultimately, the usefulness of this paper will depend on developments in supercomputer design that cannot be predicted at this time. If it turns out that a large increase in the speed of computation can be achieved by keeping down the size of the central memory, then the algorithm of this paper will indeed be useful. Conversely, it may be of some interest to those who design large computers to know that an algorithm such as this is available to exploit a machine with a small, fast central memory. (For another such algorithm, see [1].)

In this paper, then, we describe a computational method for the solution of the incompressible Navier-Stokes equations in a cubic box with periodic boundary conditions. While a periodic box is far too simple a geometry for most applications, we have shown in previous (two-dimensional) work that the interaction of a fluid with elastic boundaries that have complicated, moving, unknown geometry can be studied effectively by embedding such structures in a periodic domain and by modeling the influence of the structure on the fluid in terms of a system of forces [2,3]. (For a similar embedding approach to a different problem, see [10].) When this approach is used, the fluid equations are solved by a subroutine that only feels the elastic structures in terms of a force-field that is defined on the computational lattice. This paper has the limited goal of describing such a Navier-Stokes subroutine in the three-dimensional case.

The Navier-Stokes solver of this paper uses disc files as the primary storage medium. The files are (conceptually) subdivided into records, each record corresponding to one plane of the computational lattice. At any given time, only a few planes of data need to reside in central memory. Thus, as stated above, central memory requirements are similar to those of a 2-D Navier-Stokes solver.

The algorithm that we use to solve the incompressible Navier-Stokes equations is Chorin's finite difference method [4,5]. In Chorin's original papers, however, the Poisson equation for the pressure was solved by an iterative method, successive overrelaxation. Here we use instead a fast Poisson solver: the Fourier-Toeplitz method [6]. As we shall show, this fast Poisson solver is particularly well-suited to the efficient use of mass storage. In fact, the entire structure of the Navier-Stokes solver that we shall describe here is based on a suggestion of Olof Windlund [7] for the use of mass storage in the context of solving Poisson's equation by the Fourier-Toeplitz method as a subroutine in a capacitance-matrix computation [10].

Equations of Motion

We consider the incompressible Navier–Stokes equations:

$$(1) \qquad \rho\left[\frac{\partial \underline{u}}{\partial t} + \underline{u} \cdot \nabla\underline{u}\right] + \nabla p = \mu\, \Delta\underline{u} + \underline{F}(\underline{x},t)$$

$$(2) \qquad \nabla \cdot \underline{u} = 0$$

In these equations ρ and μ are the (constant) density and viscosity, $\underline{u}(\underline{x},t)$ and $p(\underline{x},t)$ are the (unknown) velocity and pressure, and $\underline{F}(\underline{x},t)$ is a (given) force per unit volume applied to the fluid. The position vector is $\underline{x} = (x_1, x_2, x_3)$, and t is the time. We consider $t > 0$, and we assume that the initial velocity is given:

$$(3) \qquad \underline{u}(\underline{x},0) = \underline{u}_0(\underline{x})$$

Since the fluid is incompressible, the initial pressure cannot be specified independently: the pressure can be calculated from the velocity by solving a Poisson equation.

We assume that both the initial velocity $\underline{u}_0(\underline{x})$ and the applied force $\underline{F}(\underline{x},t)$ are periodic with period L in all three coordinate directions:

$$(4) \qquad \underline{u}_0(\underline{x} + L\underline{e}_s) = \underline{u}_0(\underline{x})$$

$$(5) \qquad \underline{F}(\underline{x} + L\underline{e}_s,t) = \underline{F}(\underline{x},t)$$

where $\underline{e}_s$, $s = 1,2,3$, are unit vectors in the coordinate directions. Moreover, we impose the same periodicity on the solution $\underline{u}(\underline{x},t)$, $p(\underline{x},t)$:

$$(6) \qquad \underline{u}(\underline{x} + L\underline{e}_s,t) = \underline{u}(\underline{x},t)$$

$$(7) \qquad p(\underline{x} + L\underline{e}_s,t) = p(\underline{x},t)$$

Once such periodicity conditions are imposed, we may restrict attention to a single period such as $0 \leqslant x_1, x_2, x_3 < L$. The domain

89

is then a cube in which opposite faces (e.g., $x_1 = 0$ and $x_1 = L$) have been identified. Such a domain is called a 3-torus. It is finite, Euclidean, and translation invariant (all points are equivalent). The phrase "periodic boundary conditions", which is often used to describe this situation, is somewhat misleading: the 3-torus has no boundaries at all.

Numerical Method

We solve the Navier-Stokes equations by means of Chorin's finite-difference method: an implicit, fractional step algorithm which is also called the projection method [4,5]. To state this method, we need certain difference operators. As before, let $\underline{e}_s$, $s = 1,2,3$, be the three unit vectors in the coordinate directions, and let $h = L/N$, where N is an integer. We define forward, backward, and central difference operators in the three space directions as follows:

(8)
$$(D_s^+ \phi)(\underline{x}) = (\phi(\underline{x}+h\underline{e}_s) - \phi(\underline{x}))/h$$

(9)
$$(D_s^- \phi)(\underline{x}) = (\phi(\underline{x})-\phi(\underline{x}-h\underline{e}_s))/h$$

(10)
$$(D_s^0 \phi)(\underline{x}) = (\phi(\underline{x}+h\underline{e}_s) - \phi(\underline{x}-h\underline{e}_s))/(2h)$$

From the central difference operator, we construct a discrete gradient and divergence:

(11)
$$\underline{G}\phi = (D_1^0 \phi, D_2^0 \phi, D_3^0 \phi)$$

(12)
$$D\underline{u} = D_1^0 u_1 + D_2^0 u_2 + D_3^0 u_3$$

For future reference, it is useful to note that

(13)
$$(D_s^+ D_s^- \phi)(\underline{x}) = (\phi(\underline{x}+h\underline{e}_s) + \phi(\underline{x}-h\underline{e}_s) - 2\phi(\underline{x}))/h^2$$

(14)
$$(D\underline{G}\phi)(\underline{x}) = \sum_{s=1}^{3} (\phi(\underline{x}-2h\underline{e}_s) + \phi(\underline{x}-2h\underline{e}_s) - 2\phi(\underline{x}))/(2h)^2$$

In terms of these operators, Chorin's finite difference method may be stated as follows. Given $\underline{u}^n \sim \underline{u}(\ ,n\Delta t)$ and $\underline{F}^n \sim \underline{F}(\ ,n\Delta t)$, we compute $\underline{u}^{n+1}$, p^{n+1} as follows. First let

(15)
$$\underline{u}^{n,0} = \underline{u}^n + \frac{\Delta t}{\rho} \underline{F}^n$$

Next, solve successively the following linear systems for the auxiliary velocity fields $\underline{u}^{n,s}$, $s = 1,2,3$:

(16)
$$\rho \left[\frac{\underline{u}^{n,s} - \underline{u}^{n,s-1}}{\Delta t} + u_s^n D_s^0 \underline{u}^{n,s} \right] = \mu D_s^+ D_s^- \underline{u}^{n,s}$$

where $u_s^n = \underline{u}^n \cdot \underline{e}_s$. Finally, solve the following linear system for $\underline{u}^{n+1}$, p^{n+1}:

(17)
$$\rho \left[\frac{\underline{u}^{n+1} - \underline{u}^{n,3}}{\Delta t} \right] + \underline{G}p^{n+1} = 0$$

(18)
$$D\underline{u}^{n+1} = 0$$

In this method, the fluid velocity is updated in stages. First the external force acts on the fluid (Eq. 15). Next the convection and viscous forces are applied (Eq. 16). This is broken down into three separate steps, however, ($s = 1,2,3$): In step s, only those terms that arise from s-differences are applied. [This should not be confused with the different *components* of force. In each step, all three components are involved.] In Eq. 16, note that the convection velocity is always $\underline{u}^n$. This velocity field is known (from the previous time step), so the equations for $\underline{u}^{n,s}$ are linear. Moreover, $\underline{u}^n$ satisfies $D\underline{u}^n = 0$, which is not true of $\underline{u}^{n,s}$, $s = 0,1,2,3$. Finally, in Eq. (17), the influence of the pressure on the fluid velocity is computed. The pressure chosen must be such that $D\underline{u}^{n+1} = 0$ (Eq. 18). Thus Eqs. (17–18) form a linear system to be solved simultaneously for $\underline{u}^{n+1}$, p^{n+1}.

Solution of the Difference Equations

To complete the statement of the numerical method, we have to state the algorithm that is used to solve the implicit finite–difference equations. We begin by rewriting Eq. (16) in the form

$$(19) \qquad \{I + \Delta t(u_s^n D_s^0 - \frac{\mu}{\rho} D_s^+ D_s^-)\} \, \underline{u}^{n,s} = \underline{u}^{n,s-1}$$

These equations may be written out explicitly as a (periodic) tridiagonal system:

$$(20) \qquad -A(\underline{x})\underline{U}(\underline{x}-h\underline{e}_s) + B\underline{U}(\underline{x}) - C(\underline{x})\underline{U}(\underline{x}+h\underline{e}_s) = \underline{W}(\underline{x})$$

where $\underline{U} = \underline{u}^{n,s}$, $\underline{W} = \underline{u}^{n,s-1}$, and where

$$(21) \qquad A(\underline{x}) = \Delta t \left[\frac{u_s^n(\underline{x})}{2h} + \frac{\mu}{\rho h^2} \right]$$

$$(22) \qquad B = 1 + \Delta t \, \frac{2\mu}{\rho h^2}$$

$$(23) \qquad C(\underline{x}) = \Delta t \left[-\frac{u_s^n(\underline{x})}{2h} + \frac{\mu}{\rho h^2} \right]$$

For each s, Eq. (20) consists of $3N^2$ independent periodic tridiagonal systems, three on each of the N^2 lines of the computational lattice that run parallel to $\underline{e}_s$. (The 3 systems on a given line all have the same matrix.) Each of these systems is of order N.

Since we shall solve these systems without pivoting, it is important that they be diagonally dominant:

$$(24) \qquad B > |A(\underline{x})| + |C(\underline{x})|$$

for all $\underline{x}$. One way to achieve this is to insist that $A(\underline{x})$ and $C(\underline{x})$ be positive for all $\underline{x}$. In that case, we have

$$(25) \qquad |A(\underline{x})| + |C(\underline{x})| = A(\underline{x}) + C(\underline{x}) = B - 1$$

so $B > |A(\underline{x})| + |C(\underline{x})|$ as required. A further benefit of the restriction that $A(\underline{x})$ and $C(\underline{x})$ be positive is that this restriction guarantees the maximum-norm stability of Eq. (20). To see this, rewrite Eq. (20) in the form

$$(26) \qquad B\underline{U}(\underline{x}) = \underline{W}(\underline{x}) + A(\underline{x})\underline{U}(\underline{x}-h\underline{e}_s) + C(\underline{x})\underline{U}(\underline{x}+h\underline{e}_s)$$

Then, since $A(\underline{x})$ and $C(\underline{x})$ are positive, the following inequality holds at each $\underline{x}$

$$(27) \qquad B|\underline{U}(\underline{x})| \leq \|\underline{W}\|_{max} + (A(\underline{x}) + C(\underline{x}))\, \|\underline{U}\|_{max}$$

where $\| \cdot \|_{max}$ is defined by

$$(28) \qquad \|\underline{u}\|_{max} = \max_{\underline{x},\, s}\, |u_s(\underline{x})|$$

But $A(\underline{x}) + C(\underline{x}) = B - 1$. Therefore the inequality (27) may be rewritten

$$(29) \qquad B|\underline{U}(\underline{x})| \leq \|\underline{W}\|_{max} + (B - 1)\, \|\underline{U}\|_{max}$$

Taking the maximum over $\underline{x}$, s of both sides we get

$$(30) \qquad B\|\underline{U}\|_{max} \leq \|\underline{W}\|_{max} + (B - 1)\, \|\underline{U}\|_{max}$$

which is equivalent to the required stability property:

$$(31) \qquad \|\underline{U}\|_{max} \leq \|\underline{W}\|_{max}$$

The restriction that $A(\underline{x})$ and $C(\underline{x})$ be positive for all $\underline{x}$ amounts to a restriction on the Reynolds number in terms of N, the number of points in each direction on the computational lattice. To see this, note that $A(\underline{x})$ and $C(\underline{x})$ are positive for all $\underline{x}$ if and only if

(32)
$$\frac{\|\underline{u}^n\|_{max}}{2h} < \frac{\mu}{\rho h^2}$$

Multiplying both sides of this inequality by $L/h = N$ and rearranging the result, we get

(33)
$$R = \frac{\rho L \|\underline{u}^n\|_{max}}{\mu} < 2N$$

where R is the Reynolds number (on the nth time step).

We now show that the periodic tridiagonal system (Eq. 20) can be reduced to a pair of nonperiodic systems. First, we write Eq. (20) in the form

(34)
$$-A_j\underline{U}_{j-1} + B_j\underline{U}_j - C_j\underline{U}_{j+1} = \underline{W}_j$$

where $j = 0,...,N-1$ and where the subscript arithmetic is modulo N. Let $\underline{U}^{(0)}$ and ϕ be defined as solutions of the following (nonperiodic) problems.

(35)
$$-A_j\underline{U}_{j-1}^{(0)} + B_j\underline{U}_j^{(0)} - C_j\underline{U}_{j+1}^{(0)} = \underline{W}_j$$

(36)
$$\underline{U}_0^{(0)} = \underline{U}_N^{(0)} = 0$$

(37)
$$-A_j\phi_{j-1} + B_j\phi_j - C_j\phi_{j+1} = 0$$

(38)
$$\phi_0 = \phi_N = 1$$

where $j = 1,...,N-1$. Note that $\underline{U}^{(0)}$ has zero boundary conditions but nonzero right-hand side but that ϕ has nonzero boundary conditions and zero right-hand side. Also, ϕ is a scalar while $\underline{U}^{(0)}$ is a vector. The algorithm for solving these systems will be stated below. First, we show how to find $\underline{U}$ given $\underline{U}^{(0)}$ and ϕ.

Let $\underline{\lambda}$ be an arbitrary 3-vector. Then it is easy to see that any vector $\underline{U}$ of the form

95

(39) $$\underline{U}_j = \underline{U}_j^{(0)} + \lambda \, \phi_j$$

satisfies Eq. (34) for $j = 1,...,N-1$. To find λ, we use Eq. (34) with $j = 0$:

(40) $$-A_0\underline{U}_{N-1} + B_0\underline{U}_0 - C_0\underline{U}_1 = \underline{W}_0$$

Substituting (39) into this equation and solving for λ, we get

(41) $$\lambda = \frac{\underline{W}_0 + A_0\underline{U}_{N-1}^{(0)} + C_0\underline{U}_1^{(0)}}{B_0 - A_0\phi_{N-1} - C_0\phi_1}$$

Note that the property of maximum-norm stability derived above can now be used to show that the denominator in Eq. (41) is bounded from below by 1: From the stability property we may conclude that $|\phi_{N-1}| \leqslant 1$ and $|\phi_1| \leqslant 1$. Therefore (since A_0 and C_0 are positive),

(42) $$B_0 - A_0\phi_{N-1} - C_0\phi_1 \geqslant B_0 - (A_0 + C_0) = 1$$

The nonperiodic systems for $\underline{U}^{(0)}$ and ϕ (Eqs. 35–38) are solved by L-U factorization [9,p. 152-157,p. 165-167], and they are then combined according to Eq. (39) to obtain the periodic solution $\underline{U}$. For future reference, we write out the algorithm in detail. In the following, we use the notation

(43) $$a = \frac{\mu \, \Delta t}{\rho h^2}$$

(44) $$b = 1 + 2a$$

(45) $$u_j = \frac{\Delta t \, u_s^n(jh)}{2h}$$

(see Eqs. 21–23). The algorithm for computing $(\underline{U}^{(0)}, \phi)$ and then $\underline{U}$ may be stated as follows:

Procedure: USOLVE

Initialize: $C_1 := (-u_1+a)/b$, $\underline{U}_1 := \underline{U}_1/b$, $\phi_1 := (u_1+a)/b$

For j = 2,...,N-1, step +1, do:

$$A := u_j + a$$
$$B := b - A * C_{j-1}$$
$$C_j := (-u_j+a)/B$$
$$\underline{U}_j := (\underline{U}_j + A * \underline{U}_{j-1})/B$$
$$\phi_j := A * \phi_{j-1}/B$$

end loop.

Initialize: $\phi_{N-1} := \phi_{N-1} + C_{N-1}$

For j = N-2,...,1, step -1, do:

$$\underline{U}_j := \underline{U}_j + C_j * \underline{U}_{j+1}$$
$$\phi_j := \phi_j + C_j * \phi_{j+1}$$

end loop.

Initialize: $A := u_0+a$, $C := -u_0+a$

$$\underline{U}_0 := (\underline{U}_0 + A*\underline{U}_{N-1} + C*\underline{U}_1)/(b - A*\phi_{N-1} - C*\phi_1)$$

For j = 1,...,N-1, step +1, do:

$$\underline{U}_j := \underline{U}_j + \phi_j * \underline{U}_0$$

end loop.

End procedure.

Note that the extra work in this algorithm associated with the auxiliary variable ϕ is rather modest, since the coefficients are the same for ϕ and $\underline{U}$ and since ϕ has only one component while there are three components of $\underline{U}$.

We now consider the solution of Eqs. (17-18). We shall reduce this system to a Poisson equation for the pressure and solve the Poisson equation by the Fourier-Toeplitz method [6]. The first step in this method is to take the Fourier transform in the x_1 and x_2 directions. This reduces the Poisson equation to a collection of uncoupled periodic tridiagonal systems in the x_3 direction. These periodic tridiagonal systems are reduced to nonperiodic systems and solved as above.

To obtain the Poisson equation, we apply the divergence operator D to both sides of Eq. 17. The result (using Eq. 18) is

$$\text{(46)} \qquad\qquad DGp^{n+1} = \frac{\rho}{\Delta t} \, D\underline{u}^{n,3}$$

(Note that $D\underline{u}^{n,3} = O(\Delta t)$, so $p^{n+1} = O(1)$.) Writing out the left-hand side of Eq. (46), we find

$$\text{(47)} \qquad\qquad \sum_{s=1}^{3} \left(P(\underline{x}+2h\underline{e}_s) + P(\underline{x}-2h\underline{e}_s) - 2P(\underline{x}) \right) = Q(\underline{x})$$

where $P = p^{n+1}$ and $Q = (2h)^2 \rho \, D\underline{u}^{n,3}/\Delta t$.

The next step is to apply the discrete Fourier transform (in x_1, x_2) and thereby reduce Eq. (47) to tridiagonal form (in x_3). Now the Fourier transform is simplest in its complex form, but the variables in Eq. (47) are real. Moreover, the coupling in Eq. (47) is staggered: the point $\underline{x}$ is coupled to $\underline{x} \pm 2h\underline{e}_s$, $s = 1,2,3$. In particular, the variables on the even planes $x_3 = 2jh$ are not coupled at all to the variables on the odd planes $x_3 = (2j+1)h$. (We assume here that the total number of planes N is even.) These considerations suggest that we consider the planes in pairs and that we think of the data on each pair of planes as an array of complex numbers: Let

(48)
$$\tilde{P}(\underline{x}) = P(\underline{x}) + iP(\underline{x} + h\underline{e}_3)$$

(49)
$$\tilde{Q}(\underline{x}) = Q(\underline{x}) + iQ(\underline{x} + h\underline{e}_3)$$

To find the equations satisfied by $\tilde{P}$, $\tilde{Q}$, substitute $\underline{x} + h\underline{e}_3$ for $\underline{x}$ in Eq. (47), multiply both sides of the resulting equation by i, and add the result to Eq. (47). This procedure yields a complex equation with the same form as Eq. (47):

(50)
$$\sum_{s=1}^{3} (\tilde{P}(\underline{x} + 2h\underline{e}_s) + \tilde{P}(\underline{x} - 2h\underline{e}_s) - 2\tilde{P}(\underline{x})) = \tilde{Q}(\underline{x})$$

Note that the restriction of Eq. (50) to the even planes $x_3 = 2jh$ contains all of the information that was originally present in both the odd and even planes of Eq. (47). This information can be extracted by taking the real and imaginary pairs.

We now take the discrete Fourier transform of Eq. (50). Let

(51)
$$\tilde{P}(\underline{x}) = \sum_{k_1=0}^{N-1} \sum_{k_2=0}^{N-1} \hat{P}(k_1,k_2;x_3) \exp(i \frac{2\pi}{L} (k_1 x_1 + k_2 x_2))$$

(52)
$$\tilde{Q}(\underline{x}) = \sum_{k_1=0}^{N-1} \sum_{k_2=0}^{N-1} \hat{Q}(k_1,k_2;x_3) \exp(i \frac{2\pi}{L} (k_1 x_1 + k_2 x_2))$$

Substituting these expressions into Eq. (50) and collecting the coefficients of $\exp(i \frac{2\pi}{L} (k_1 x_1 + k_2 x_2))$, we find

(53)
$$\sum_{s=1}^{2} (\exp(i \frac{2\pi}{L} k_s 2h) + \exp(-i \frac{2\pi}{L} k_s 2h) - 2) \hat{P}(k_1,k_2;x_3)$$

$$+ \hat{P}(k_1,k_2;x_3+2h) + \hat{P}(k_1,k_2;x_3-2h)$$

$$- 2\hat{P}(k_1,k_2;x_3) = \hat{Q}(k_1,k_2;x_3)$$

This is a collection of periodic tridiagonal systems of the form

(54) $-\hat{A}\ \hat{P}(k_1,k_2;x_3-2h) + \hat{B}(k_1,k_2)\ \hat{P}(k_1,k_2;x_3) - \hat{C}\ \hat{P}(k_1,k_2;x_3+2h)$

$$= -\hat{Q}(k_1,k_2;x_3)$$

where

(55) $$\hat{A} = \hat{C} = 1$$

(56) $\hat{B}(k_1,k_2) = 2 - 2 \sum_{s=1}^{2} (\cos \frac{2\pi}{L} k_s\ 2h - 1)$

$$= 2 (1 + 2 \sum_{s=1}^{2} \sin^2 \frac{2\pi}{N} k_s)$$

As remarked above, it is sufficient to consider the even planes $x_3 = 2jh$. Once $\hat{P}$ has been computed on these planes, we can apply the inverse Fourier transform (Eq. 51) to obtain $\tilde{P}$. Then

(57) $$P(x_1,x_2,2jh) = \text{Re}\ \tilde{P}(x_1,x_2,2jh)$$

(58) $$P(x_1,x_2,(2j+1)h) = \text{Im}\ \tilde{P}(x_1,x_2,2jh)$$

The advantage of combining the even and odd planes in this way is that we retain the simplicity of the complex Fourier transform without wasting any storage. The complex Fourier transform is computed two planes at a time with the even plane treated as the real part of the data and the odd plane as the imaginary part of the data. [This is not how FORTRAN stores complex numbers, so we have to write out the complex arithmetic in terms of real and imaginary parts.] A peculiar feature of this algorithm is that it allows data from the even and odd planes to interact, even though the problems on the even and odd planes are uncoupled. This "mixup" is sorted out when the inverse Fourier transform is computed.

The periodicity of Eq. (54) is removed by the same method used to remove the periodicity in Eq. (20). Let $\tilde{P}^{(0)}$ and $\tilde{\phi}$ be defined as the solutions of

(59) $-\hat{P}^{(0)}(k_1,k_2;x_3-2h) + \hat{B}(k_1,k_2)\hat{P}^{(0)}(k_1,k_2;x_3)$

$-\hat{P}^{(0)}(k_1,k_2;x_3+2h) = -\hat{Q}(k_1,k_2,x_3)$

(60) $-\hat{\phi}(k_1,k_2;x_3-2h) + \hat{B}(k_1,k_2)\hat{\phi}(k_1,k_2;x_3)$

$-\hat{\phi}(k_1,k_2;x_3+2h) = 0$

where $x_3 = 2jh$, $j = 1,2,...,(N/2)-1$, and where

(61) $\hat{P}^{(0)}(k_1,k_2;0) = \hat{P}^{(0)}(k_1,k_2;L) = 0$

(62) $\hat{\phi}(k_1,k_2;0) = \hat{\phi}(k_1,k_2;L) = 1$

Then $\hat{P}$ can be expressed as a linear combination of $\hat{P}^{(0)}$ and $\hat{\phi}$:

(63) $\hat{P}(k_1,k_2;x_3) = \hat{P}^{(0)}(k_1,k_2;x_3) + \lambda(k_1,k_2)\hat{\phi}(k_1,k_2;x_3)$

Note that λ here is a (complex) scalar. The equation for λ is obtained by substituting Eq. (63) into Eq. (54) with $x_3 = 0$. The result is:

(64) $\lambda(k_1,k_2) = \dfrac{-\hat{Q}(k_1,k_2;0)+\hat{P}^{(0)}(k_1,k_2;L-2h)+\hat{P}^{(0)}(k_1,k_2;2h)}{\hat{B}(k_1,k_2)-\hat{\phi}(k_1,k_2;L-2h)-\hat{\phi}(k_1,k_2;2h)}$

We now investigate the denominator of Eq. (64). First, we note that, since $\hat{B}(k_1,k_2) \geq 2$, the following inequality holds at all interior points:

(65) $\hat{\phi}(k_1,k_2;x_3) \leq \frac{1}{2}[\hat{\phi}(k_1,k_2;x_3+2h) + \hat{\phi}(k_1,k_2;x_3-2h)]$

It follows that $\hat{\phi}$ has no interior maxima, and hence that

(66) $\hat{\phi}(k_1,k_2;x_3) \leq 1$

There are now two cases to consider. First, suppose that

101

$(k_1,k_2) \notin S$, where

(67) $$S = \{(0,0),(0,N/2),(N/2,0),(N/2,N/2)\}$$

Then

(68) $$\hat{B}(k_1,k_2) \geq 2 + 4 \sin^2(2\pi/N)$$

Thus, for $(k_1,k_2) \notin S$, the denominator in Eq. (64) is boudned from below by $4 \sin^2(2\pi/N)$. Although this is $O(N^{-2})$, it is still a moderate number (compared to machine precision) for reasonable values of N. For example, if $N = 64$, $2\pi/N = 0.1$, and $4 \sin^2(2\pi/N) = 0.04$. Even with the fantastic resolution of $N = 1024$ we would still get the moderate number $4 \sin^2(2\pi/N) = 0.00015$. In short, for $(k_1,k_2) \notin S$, we do not have to worry about the denominator in Eq. (64) approaching zero.

For $(k_1,k_2) \in S$, on the other hand, we have $\hat{B}(k_1,k_2) = 2$ and $\hat{\phi}(k_1,k_2;x_3) = 1$ for all x_3. Thus, the denominator in Eq. (64) is exactly zero. This means that Eq. (54) is singular for these four values of (k_1,k_2) and that the equation has solutions if and only if the numerator in Eq. (64) is also zero. This is equivalent to the condition

(69) $$\sum_{j=0}^{N/2-1} \hat{Q}(k_1,k_2;2jh) = 0, \qquad (k_1,k_2) \in S$$

To demonstrate the equivalence, sum Eq. (59) with $\hat{B} = 2$ over all (even) interior planes and recall the boundary conditions, Eq. (61). [It can also be shown directly that Eq. (69) is a necessary condition for the existence of solutions to Eq. (54) when $(k_1,k_2) \in S$. To show this, just sum Eq. (54) with $\hat{B} = 2$ over all even planes (including $x_3 = 0$).]

Fortunately, Eq. (69) is always satisfied. It is an immediate consequence of the definition of Q as a discrete divergence of a vector field on a periodic domain (see after Eq. 47).

Thus, for $(k_1,k_2) \in S$, Eq. (64) yields $\lambda(k_1,k_2) = 0/0$, and λ is undefined. This makes sense, since Eq. (63) yields a solution to

Eq. (54) for all λ when $(k_1,k_2) \in S$ (recall that $\hat{\phi} = 1$ for $(k_1,k_2) \in S$). Moreover, it is easy to check that the choice of λ for $(k_1,k_2) \in S$ has no effect on the discrete gradient of P and hence on the computed velocity field $\underline{u}^{n+1}$. We fix λ by imposing the same condition on $\hat{P}$ that $\hat{Q}$ automatically satisfies:

$$(70) \qquad \sum_{j=0}^{N/2-1} \hat{P}(k_1,k_2;2jh) = 0, \qquad (k_1,k_2) \in S$$

(compare Eq. 69). These four complex conditions are equivalent to setting the average of P equal to zero on each of the 8 "chains" of lattice points $C(x_1,x_2,x_3) = \{(x_1+2j_1h,x_2+2j_2h,x_3+2j_3h)\}$, where j_1,j_2,j_3 are arbitrary integers and where $x_s = 0$ or h, $s = 1,2,3$.

To find an explicit formula for λ when $(k_1,k_2) \in S$ we sum Eq. (63) over all even planes. Because of Eq. (70), the left hand side of Eq. (63) sums to zero. Moreover, since $\hat{\phi}(k_1,k_2;x_3) = 1$ for $(k_1,k_2) \in S$, we have

$$(71) \qquad \lambda(k_1,k_2) = \frac{-1}{N} \sum_{j=0}^{N/2-1} \hat{P}^{(0)}(k_1,k_2;2jh)$$

$$= \frac{-1}{N} \sum_{j=1}^{N/2-1} \hat{P}^{(0)}(k_1,k_2;2jh), \qquad (k_1,k_2) \in S$$

(In the second expression the term $j = 0$ has been dropped, since it is zero in any case.)

To complete the description of the solution for the pressure, we need to specify how Eqs. (59-62) are solved for $\hat{P}^{(0)}$ and $\hat{\phi}$. These nonperiodic tridiagonal systems are symmetric and positive definite (even when $(k_1,k_2) \in S$); they are solved by LU factorization without pivoting.

For future reference, we state the algorithm that computes $\hat{P}$ from $(-\hat{Q})$. The algorithm uses the same storage, designated $\hat{P}$, for both quantities. In the following, $b(k_1,k_2)$ is the function given by the right-hand side of Eq. (56).

Procedure: PSOLVE

103

<u>Initialize:</u> $C_2 := 1/b(k_1,k_2)$, $\hat{P}_2 := \hat{P}_2/b(k_1,k_2)$, $\hat{\phi}_2 := 1/b(k_1,k_2)$

For j = 4,...,N-2, step +2, do:

$$B_j := b(k_1,k_2) - C_{j-2}$$
$$C_j := B_j^{-1}$$
$$\hat{P}_j := (\hat{P}_j + \hat{P}_{j-2})/B_j$$
$$\hat{\phi}_j := \hat{\phi}_{j-2}/B_j$$

end loop.

<u>Initialize:</u> $\hat{\phi}_{N-2} := \hat{\phi}_{N-2} + C_{N-2}$

If $(k_1,k_2) \in S$ then PSUM $:= \hat{P}_{N-2}$

For j = N-4,...,2, step -2, do:

$$\hat{P}_j := \hat{P}_j + C_j * \hat{P}_{j+2}$$
$$\hat{\phi}_j := \hat{\phi}_j + C_j * \hat{\phi}_{j+2}$$
If $(k_1,k_2) \in S$ then PUM $:=$ PSUM $+ \hat{P}_j$

end loop.

<u>Initialize:</u>

If $(k_1,k_2) \in S$, then
$$\hat{P}_0 := -PSUM/N$$
else,
$$\hat{P}_0 := (-\hat{P}_0 + \hat{P}_{N-2} + \hat{P}_2)/(b(k_1,k_2) - \hat{\phi}_{N-2} - \hat{\phi}_2)$$
end if.

For j = 2,...,N-2, step +2, do:

$$\hat{P}_j = \hat{P}_j + \hat{P}_0 * \hat{\phi}_j$$

end loop.

104

End procedure.

Although we shall discuss implementation in greater detail below, a few remarks concerning the implementation of PSOLVE are in order here. First, we note that several of the quantities appearing in PSOLVE are the same at every time step. These quantities, B_j, C_j, and $\hat{\phi}$, can be computed once and for all and stored. Similarly, the denominator in the expression for $\hat{P}_0$ (note that $\hat{P}_0 \equiv \lambda$) is a fixed function of (k_1, k_2) which can therefore be precomputed and stored. Second, we note that this algorithm is not actually used separately for each (k_1, k_2). Instead, all N^2 values of (k_1, k_2) are done "at once". That is, each statement should be read as a loop over (k_1, k_2). Finally, we remark that the algorithm can be implemented without any IF statements. The four (complex) quantities in PSUM can be initialized and updated by a routine that only looks at pairs $(k_1, k_2) \in S$. At the step in which $\hat{P}_0$ (i.e., λ) is computed, the values of $-PSUM/N$ can be inserted into the numerator array $(-\hat{P}_0 + \hat{P}_{N-2} + \hat{P}_2)$ at the appropriate locations overwriting the zeroes that are there. In the denominator array $(b(k_1, k_2) - \hat{\phi}_{N-2} - \hat{\phi}_2)$ there are also zeroes at these four locations but they can be replaced by ones. (This is done when the denominator is precomputed, see above.) Then the general formula for λ (the "else" case) can be used for all N^2 values of (k_1, k_2).

At this stage we can rewrite the numerical method (Eqs. 15–18) in terms of the computational procedures that are used to compute $(\underline{u}^{n+1}, p^{n+1})$ from $(\underline{u}^n, \underline{F}^n)$:

$$\underline{u} := \underline{u}^n + \frac{\Delta t}{\rho} \, \underline{F}^n$$

$$\underline{u} := USOLVE_1(u_1^n, \underline{u})$$

$$\underline{u} := USOLVE_2(u_2^n, \underline{u})$$

$$\underline{u} := USOLVE_3(u_3^n, \underline{u})$$

$$p := -\frac{(2h)^2 \rho}{\Delta t} \, D\underline{u}$$

$$p := FFT2D(p)$$

$$p := PSOLVE(p)$$

$$p := \text{FFT2D}^{-1}(p)$$
$$\underline{u} := \underline{u} - \frac{\Delta t}{\rho}\, G\, p$$

At this point p holds p^{n+1}, $\underline{u}$ holds $\underline{u}^{n+1}$, and the time step is complete.

Asynchronous Input/Output

In this section we discuss communication between central memory and a mass storage device (disk or SSD[*]). We shall describe a collection of FORTRAN programming techniques to streamline such communication. The aim of these techniques is to achieve maximal overlap of input, output, and computation. Such overlap is possible because the central processing unit is only needed to *initiate* a transfer of information between central memory and the disk. Once initiated, such transfer can be handled by separate input–output processors while the central processor continues with the computation. This mode of operation is called *asynchronous* i/o (input/output) because the central processor and the i/o processors are not synchronized.

It is important to note that asynchronous i/o makes possible not only the overlap of i/o with computation but also the overlap of i/o with itself. That is, input or output can proceed from or to several different files at the same time.[**] To facilitate the overlap of i/o with itself, we have used a software trick that makes it possible to read and write data from and to the "same" file at the same time. In reality, two different files are used but this fact is transparent to the user.

In this work, we use the disk as the principal storage medium. A lattice function (such as the pressure or any individual component of the velocity) is stored on the disk in a file. We think of such a file as being subdivided into N records numbered $0...N-1$. (There are no

---

[*] SSD stands for *Solid-state Storage Device* or *Solid-State Disk*. As far as the user is concerned, an SSD is just a very fast disk. In fact, *no* FORTRAN reprogramming is required when an SSD is substituted for a conventional disk; the only changes are at the level of job-control language. For this reason, we do not distinguish the two devices when describing our programming techniques. Except where an SSD is specifically mentioned, we use *disk* as a generic term that covers both possibilities.

[**] The extent to which this is really possible depends on some hardware and system-dependent features such as the number of available channels for communication between disk and central memory.

end-of-record marks, however, since *unblocked* files are used.) Each record holds the data corresponding to one horizontal (x_3 = constant) plane of the computational lattice. For variables that are updated repeatedly as the computation proceeds, we use two such disk files: one to hold the old values that are read into central memory and the other to hold the new values that are written onto the disk from central memory. We refer to these as the READ copy and the WRITE copy of the file, respectively. After a variable has been completely updated, we interchange the labels READ and WRITE so that the latest data can be found in the READ file and so that the WRITE file contains obsolete data that may be safely overwritten. [For this strategy to work, it is crucial that every file be completely rewritten when any part of it is rewritten. Fortunately, our algorithm has this property.]

To make this double representation of files transparent to the user, we assign a *symbolic unit number* (an integer) to each lattice function. Corresponding to each symbolic unit number, there are two actual FORTRAN unit numbers, one for the READ file and one for the WRITE file. These are stored in the integer array URW with dimensions URW(2, # of lattice functions). The meaning of the numbers stored in URW is as follows:

URW(1,USM) = READ unit corresponding to USYM
URW(2,USM) = WRITE unit corresponding to USYM

A subroutine SWITCH(USYM) is provided which simply interchanges the numbers stored in URW(1,USYM) and URW(2,USYM). This has the effect of interchanging the labels READ and WRITE as described above. Obviously, SWITCH is a very fast routine. In particular, it does not involve copying data from one file to another! Since the READ and WRITE files corresponding to a given symbolic unit number are actually assigned to different FORTRAN units, they are treated as different files by the system and there is no difficulty in assessing them both at the same time via asynchronous i/o.

Incidentally, there are some lattice functions that are never rewritten. Such functions are computed during initialization and stored on the disk; then they are used repeatedly without further change. For such lattice functions we again assign a symbolic unit number USYM, but we set URW(2,USYM) = URW(1,USYM). Then the READ and WRITE versions of the file are the same. [If SWITCH(USYM) should happen to be called, it would have no effect.]

Up to this point we have discussed the organization of data on the disk. We now turn to a discussion of the corresponding data structure in central memory. For each lattice function that is stored on the disk, we have a corresponding array in central memory. This array is big enough to hold 4 planes of data, and it is dimensioned as follows:

```
PARAMETER (L2NG = 6, NG = 2 ** L2NG)
PARAMETER (NSIZE = (NG+2)*NG, NB = NG+1)
PARAMETER (NW = (1 + NSIZE/512) * 512)
PARAMETER (NGY = 1 + NW/(NG+2))
DIMENSION U(0:NB, 1:NGY, 0:3)
```

The meaning of the parameters and dimensions defined above is as follows (see also Fig. 1). First, NG is the number of grid (i.e.,

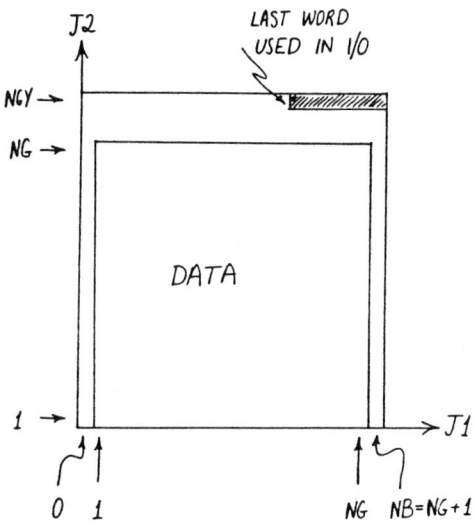

Figure 1

109

lattice) points in each direction (NG = N). In the x_1 and x_2 directions, we number the grid points from 1...NG, whereas in x_3 we use 0...(NG-1). We assume that NG is a power of 2, the power being L2NG. In the above example L2NG = 6, so NG = 64. To avoid memory-bank conflicts[*], we put borders on each array (such as U) in the x_1 direction. That is, we let I, the x_1 index, vary from 0 through NB = NG+1. These borders are also useful in the computation of divergence and gradient, see below.

No such borders are needed in the x_2 direction, so J, the x_2 index, starts from 1. We extend the array U in the x_2 direction, however, because of a restriction imposed by input/output considerations: The fastest i/o is achieved with *unblocked* files, and a read or write operation on such a file must be in integer multiples of 512 words. The number of words that we actually want to read or write at a time is NSIZE = NG * (NG+2); note that this necessarily includes the borders described above since a single read or write must refer to contiguous memory locations. Since NSIZE is not usually a multiple of 512, we find the first multiple of 512 that is greater than NSIZE. This is called NW; it is the number of words that will be transferred by each read or write statement. Finally, we have to extend the aray U in the x_2 direction so that each plane of U contains *at least* NW words. (Fortunately, it is all right if the plane contains *more* than NW words, as there is no restriction that an entire array be read or written on each i/o transfer.) Therefore

---

[*]The central memory of the Cray 1 or Cray X-MP is divided into 16 memory banks. Contiguous words of memory are stored in *successive* banks. Therefore, memory locations whose addresses differ by a multiple of 16 are physically located in the same bank. If we sweep through a two-dimensional array by varying its first subscript in steps of 1, the successive central memory requests are issued to different memory banks; this promotes fast memory access. If we do the same with the second subscript, however, and if the first array dimension is a multiple of 16, then all requests will go to the same bank; this slows down the memory access. We avoid this problem by putting borders on the array in question so that the first dimension is not a multiple of 16. (Caution: the memory organization of the Cray 2 is different; for best results the first dimension should be *odd*.)

we choose NGY, the upper limit of J, such that NGY is the smallest integer to satisfy NGY * (NG+2) > NW.

The x_3 index of the array U varies from 0 through 3. Thus the array U is big enough to hold 4 planes of data. We use the x_3 index in cyclic fashion. That is, we store plane K in U($\cdot$,$\cdot$,KX), where

$$KX = MOD4(K)^*$$

The typical use of the data structure that we have described is as follows. An algorithm is executed that sweeps through the planes K in (increasing or decreasing) sequence. At each stage of the algorithm, the following three tasks are performed in parallel:

(i) Data are read into central memory from the plane *ahead* of plane K.

(ii) Computation is performed on plane K, possibly using (but not changing) data from the plane *behind* plane K.

(iii) Data are written to the disk from the plane *behind* plane K.

As the plane index K goes from 0...(NG–1), in either order, the corresponding central-memory index KX cycles repeatedly through 0...3 in the corresponding order. A specific example that implements this approach will be given at the end of this section.

Having described the data structure on the disc and in central memory, we now proceed to a discussion of the FORTRAN routines that are used for input/output. These routines provide an interface between the applications program and the i/o routines provided by the

---

* We supply the function MOD4 which gives correct values even for moderately negative arguments, which the FORTRAN function MOD($\cdot$,4) does not. Our definition is MOD4(K) = MOD(K+16,4), which works as long as K $\geq$ –16.

Cray extensions to FORTRAN. The form of this interface follows closely that of a similar but more general-purpose interface written by C. Hsiung of Cray Research (unpublished communication). The main novel feature here is the assignment of a READ file and a WRITE file to each symbolic unit.

We shall describe our i/o interface in top-down fashion. The highest level routines will be described here in terms of their use. They will also be listed in the Appendix together with the lower-level routines that they call. The top level routines are as follows:

READIN(UNIT U, K, U, KX)

WRTOUT(UNIT U, K, U, KX)

WAITR(UNIT U)

WAITW(UNIT U)

SWITCH(UNIT U)

Here UNIT U is an integer variable that holds the symbolic unit number of the disk file in question. [The space after UNIT is a convention for the user that has no meaning to the FORTRAN compiler.] The corresponding central-memory array U is dimensioned as described above. The index K denotes the record number in the pair of files designated by UNIT U, and the index KX denotes which plane of the central-memory array U is to be used. Ordinarily, $KX = MOD4(K)$. A call to READIN initiates the transfer of record K from the READ file of UNIT U into the central memory array $U(\cdot,\cdot,KX)$. Similarly a call to WRTOUT initiates the transfer of data from the central-memory array $U(\cdot,\cdot,KX)$ onto record K of the WRITE file of UNIT U. Once the transfer of data is initiated in either direction, the central processing unit returns to the task of executing the program while the information transfer proceeds in parallel.

The routine WAITR(UNIT U) suspends operation of the program until the READ file of UNIT U is no longer busy. Similarly, WAITW(UNIT U) suspends operation of the program until the WRITE file of UNIT U is no longer busy. Note that calls to WAITR and WAITW only suspend execution of the program by the central processing unit; they do not suspend any i/o operations that happen to be in progress. It is also worth mentioning the low cost of a (superfluous) call to WAITR or WAITW when the file in question is not actually busy. In such a case, the wait routine returns immediately.

The last routine, SWITCH(UNIT U), simply interchanges the roles of the READ and WRITE files corresponding to UNIT U. This routine is used after the file has been completely updated.

Some additional features of the routines READIN and WRTOUT are as follows. First, both routines check that the record number K is in range $(0 \leqslant K \leqslant (NG-1))$. If not, the READIN and WRTOUT routines RETURN immediately without performing any i/o. This avoids some IF statements that would otherwise clutter the programs that call READIN and WRTOUT. Second, both routines wait to make sure that the corresponding file is available (not busy) *before* initiating any i/o. That is, READIN calls WAITR(UNIT U) and WRTOUT calls WAITW(UNIT U). Note that these calls are made before, not after, the initiation of i/o transfer. Note further that READIN does not wait for a WRITE on UNIT U to finish nor does WRTOUT wait for a READ on UNIT U to finish. Thus, as explained above, we can perform a READ and a WRITE on the same *symbolic* unit at the same time.

Finally, the routines READIN and WRTOUT also refer to a logical variable SYNCON (which is stored in a common block). When SYNCON = .TRUE., these routines call WAITR and WAITW, respectively, not only before but also *after* the initiation of i/o transfer. This has the effect of switching the program to synchronous i/o or as long as SYNCON = .TRUE., since each i/o operation must be *completed* before control returns to the calling program. The switch SYNCON is an extremely useful debugging aid. In particular, it is good for detecting the type of error in which a call to WAITR or WAITW was needed at some point in the program but not supplied. The end results of the computation should be identical with SYNCON = .TRUE. or with SYNCON = .FALSE.; if they are not there is a bug of the type described above, i.e., there is a failure to synchronize the program at some point where synchronization is

needed. The switch SYNCON is also useful for timing. Comparison of
the timing with SYNCON = .TRUE. versus SYNCON = .FALSE. shows
how much is gained through the use of asynchronous i/o. Finally
SYNCON can be temporarily set equal to .TRUE. if there are some
parts of the program that require synchronous i/o. We use SYNCON
in this way during initialization.

We conclude this section with an example that illustrates the
use of all of the apparatus that we have described above. We give
first a simple in-core computation which requires 3D central memory,
and then we give the corresponding out-of-core version which requires
only 2D central memory. The in-core statement of the computation to
be performed is as follows:

```
      DO 1 K=1,NG-1
      DO 1 J=1,NG
      DO 1 I=1,NG
1        U(I,J,K) = U(I,J,K) + A(I,J,K) * U(I,J,K-1)
```

The out-of-core implementation of this computation using asynchronous
i/o is the following (see also Fig. 2):

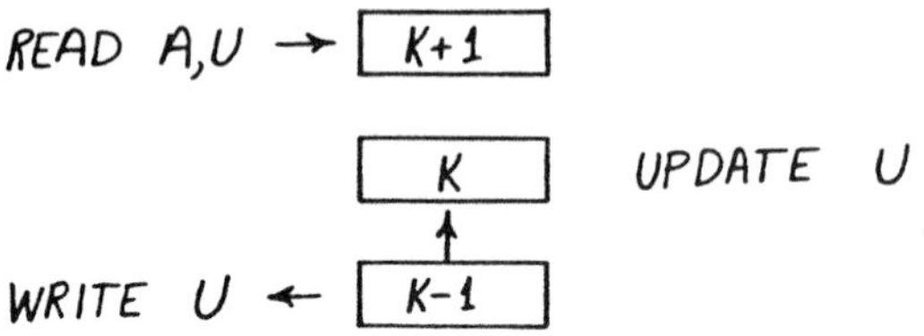

Figure 2

```
      DO 1 K=-1,NG
      KXP1 = MOD4(K+1)
      KX   = MOD4(K)
      KXM1 = MOD4(K-1)
      CALL READIN(UNIT U,K+1,U,KXP1)
      CALL READIN(UNIT A,K+1,A,KXP1)
      CALL WRTOUT(UNIT U,K-1,U,KXM1)
```

114

```
          IF(1 .LE. K) .AND. (K .LE. NG-1)) THEN
                DO 2 J = 1,NG
                DO 2 I = 1,NG
    2           U(I,J,KX) = U(I,J,KX) + A(I,J,KX) * U(I,J,KXM1)
          END IF
          CALL WAITR(UNIT U)
          CALL WAITR(UNIT A)
          CALL WAITW(UNIT U)
    1     CONTINUE
          CALL SWITCH(UNIT U)
```

There are certain aspects of this program that require comment. First, note that K goes from -1 through NG, although computation is only performed on $1 \leqslant K \leqslant (NG-1)$. When $K = -1$, the data from plane 0 are read in; when $K = 0$, the data from plane 1 are read in. Then the computation is ready to start. Similarly, when $K = NG$, the last results corresponding to plane NG-1 are written to the disk. The READIN statements corresponding to $K+1 = NG$ and $K+1 = NG+1$ are automatically ignored (see above) as are the WRTOUT statements corresponding to $K-1 = -2$ and -1.

Next, the reader may have noticed that for this particular example only 3 central-memory planes are needed in each array. There is no harm in having 4, however, and we shall need 4 in our actual use of the strategy that is illustrated here.

The next point to notice is an asymmetry between reading and writing. If a READ is in progress into a central-memory array, then that array can neither be *used* nor *changed* with reliable results. If it is used, there is no way to know whether the data used are the values from before or after the READ. If it is changed, then the change may subsequently be overwritten by the READ statement and hence undone. While a WRITE is in progress, the data cannot be *changed* with reliable results, but they can be *used*, since the WRITE operation does not alter the central-memory values. We exploit this in the foregoing by writing out plane K-1 while simultaneously making use of the data on this plane in the computation.

115

Finally, the reader may be puzzled by the calls to WAITR and WAITW at the end of each passage through the loop. These seem superfluous because there are similar calls built into the READIN and WRTOUT routines as described above. The function of this group of calls is to make sure that *all* i/o from step K of the loop is complete before step (K+1) is entered at all. In the present example, this precaution is not necessary, but we have found similar-looking programs which are actually incorrect without such synchronization. (To generate such an example, replace MOD4 by MOD3 in the foregoing, and think about what happens if the READ statement of step K+1 through the loop starts before the WRITE statement of step K is complete.) It therefore seems prudent to include these synchronization statements as a matter of routine.

Vectorization

The computational algorithm of this paper is structured in such a way that all inner loops are successfully vectorized by the Cray FORTRAN compiler. Moreover, when NG = 64 (or a multiple thereof), the vector registers, which have length 64, are fully utilized. (Through the use of parameter statements, we let the compiler know the value of NG.)

Vectorization is achieved throughout by a very simple device, which was first pointed out to us by Loyce Adams in the context of solving a large number of tridiagonal systems. In every case, the computations we perform consist of a large number of applications of the same algorithm to different data sets. In this situation, vectorization can always be achieved by letting the inner-loop index (or indices) run over the different data sets while the outer-loop index (or indices) run through the algorithm itself. The applications of this strategy in our code are as follows.

1.   Tridiagonal systems:

(a) x_3 direction: To solve $(NG)^2$ periodic tridiagonal systems in the x_3 direction, one system on each vertical column of the computational lattice, we let the inner loops run through x_1 and x_2, while the outer loop runs through x_3. [Note that this choice of inner loops is also required by our i/o strategy, described above.]

(b) x_1 direction: On a given plane, we solve NG periodic tridiagonal systems in the x_1 direction, one for each of the lines x_2 = constant. The inner loops run through the values of x_2.

(c) x_2 direction: Same as the foregoing but with x_1 and x_2 interchanged. The inner loops run through the values of x_1.[*]

2.   Fourier transform. On each pair of planes, the two-dimensional Fourier transform is computed by first taking the transform in the x_1 direction and then in the x_2 direction. Thus, we vectorize the computation as follows:

(a)   To compute NG separate Fourier transforms in the x_1 direction (along the lines x_2 = constant) on a given pair of planes, we let the inner loops run through the values of x_2.

(b)   To compute NG separate Fourier transforms in the x_2 direction (along the lines x_1 = constant) on a given pair of planes, we let the inner loops run through the values of x_1.[**]

---

[*] The *stride* through memory is different in cases (1b) and (1c). In case (1c) it is 1, while in case (1b) it is equal to the first dimension of the arrays in question. To avoid memory bank conflicts on the Cray 1 or Cray X-MP, this dimension should not be a multiple of 16. As described above, we put borders on our arrays to make the first dimension NG+2. The success of this strategy is shown by timing the routines that carry out (1b) and (1c). The times are virtually identical.

[**] Again, the stride through memory is different in cases (2a) and (2b), but our choice of the first array dimension avoids memory-bank conflicts.

A nice feature of the strategy that we have just outlined is that it leads to a simple recipe for generating vectorized code: First, write a program for a single application of the algorithm in question. Then, take each statement (or block of contiguous statements) in the program that refers to an item of data, and replace that statement by a loop over the different datasets to which the algorithm will be applied. [It is also necessary, of course, to add one dimension to each data variable. The index of this extra dimension becomes the index of the inner loops.]

In the case of the Fourier transform, for example, we started with the complex FFT code of [8] as reproduced by Dahlquist and Björk [9], p. 416. The complex, one-dimensional array A(I) was replaced by a pair of real, two-dimensional arrays A(I,J), B(I,J) in which A holds the real part and B the imaginary part of the complex data. The original FFT code has two main parts: a bit-reversal permutation followed by the arithmetic of the Fourier transform itself. The central lines of the bit-reversal permutation in the original code are as follows:[*]

```
T    = A(I1)
A(I1) = A(I2)
A(I2) = T
```

In our code for the x_1 direction, the foregoing lines are replaced by the following loops:

---

[*] The notation has been changed slightly for greater clarity.

```
        DO 91 J = 1,N

        TA1(J)  = A(I1,J)
        TB1(J)  = B(I1,J)
        TA2(J)  = A(I2,J)
   91   TB2(J)  = B(I2,J)

        DO 92 J = 1,N

        A(I1,J) = TA2(J)
        B(I1,J) = TB2(J)
        A(I2,J) = TA1(J)
   92   B(I2,J) = TB1(J)
```

these loops obviously vectorize. The code for the x_2 direction has the same structure, but loops run through the first index I, and the interchange is made between J1 and J2.

Similarly, the key arithmetic operations of the one-dimensional, complex FFT are as follows:

```
   T     = A(I1) + U * A(I2)
   A(I2) = A(I1) - U * A(I2)
   A(I1) = T
```

As before, these are replaced by the following loops:

```
        DO 93 J = 1,N

        TA1(J)  = A(I1,J)
        TB1(J)  = B(I1,J)
        TA2(J)  = UA * A(I2,J) - UB * B(I2,J)
   93   TB2(J)  = UA * B(I2,J) + UB * A(I2,J)

        DO 94 J = 1,N

        A(I1,J) = TA1(J) + TA2(J)
```

$$B(I1,J) = TB1(J) + TB2(J)$$
$$A(I2,J) = TA1(J) - TA2(J)$$
94 $\quad B(I2,J) = TB1(J) - TB2(J)$

These loops vectorize, and the corresponding code for the x_2 direction has the same structure: the loop index is I in that case and the arithmetic involves J1 and J2.

A different aspect of vectorization concerns the periodic character of the domain. This comes up, for example, when we compute the divergence of the periodic lattice vector (U,V,W). Suppose U(I,J), V(I,J) are the x_1 and x_2 components of velocity on a particular plane $x_3 = $ constant, and let WP1(I,J), WM1(I,J) be the x_3 components on the plane above and below, respectively. Then the divergence on the plane in question is given by

$$D(I,J) = ((U(I+1,J) - U(I-1,J))$$
$$+ (V(I,J+1) - V(I,J-1))$$
$$+ (WP1(I,J) - WM1(I,J)))$$

In this formula, though, the subscript arithmetic has the following special interpretation on account of the periodic character of the domain:

$$NG + 1 = 1$$
$$1 - 1 = NG$$

We handle this differently in the I and J directions because I will be the inner loop and because only the inner loop will vectorize in any case. For the J direction we use table lookup: During initialization we define two arrays NP1, NM1 such that

$$NP1(J) = J+1, \quad J = 1,...,NG-1$$
$$NP1(NG) = 1$$
$$NM1(J) = J-1, \quad J = 2,...,NG$$
$$NM1(1) = NG$$

121

In the I direction, on the other hand, we make use of the borders on the central-memory arrays. [These borders were introduced above to avoid memory-bank conflicts. Here we find an additional use for them.] That is, we explicitly copy the values of U on I = NG into I = 0 and the values of U on I = 1 into I = NG+1. This leads to the following code:

```
DO1 J = 1,NG
JP1 = NP1(J)
JM1 = NM1(J)
U(0,J) = U(NG,J)
U(NG+1,J) = U(1,J)
DO 1 I = 1,NG
D(I,J) = ((U(I+1,J) - U(I-1,J))
        + (V(I,JP1) - V(I,JM1))
        + (WP1(I,J) - WM1(I,J)))
1       CONTINUE
```

The inner loop vectorizes, and there is no need for special cases at the edges of the array. The same approach is used in the computation of the gradient of the pressure.

Subroutine FLUID

In this section, we outline the subroutine that solves the Navier-Stokes equations using the techniques described above. The main point of this section is to explain how the different operations are intertwined in such a way as to avoid all unnecessary i/o. That is, once a given plane of data is in central memory, we perform as much computation on this plane as possible before returning results to the disc.

The principal input to subroutine FLUID consists of the disc files:

$$(UNIT\ UN,\ UNIT\ VN,\ UNIT\ WN)$$

$$(UNIT\ U\ ,\ UNIT\ V\ ,\ UNIT\ W\ )$$

The first triple of files holds the velocity $\underline{u}^n$ and the second triple holds $\underline{u}^n + (\Delta t/\rho)\underline{F}^n$. On output, these two triples of files will both hold the vector $\underline{u}^{n+1}$. (An additional output is the disc file UNIT P, which holds p^{n+1}.)

The computation proceeds in 5 main loops numbered 100, 200, etc., in the course of which we sweep UP, DOWN, UP, DOWN, UP through the data. The computations performed in these different loops are as follows:

loop 100 $(K\uparrow)$:USOLVE$_3$ (forward elimination sweep)
loop 200 $(K\downarrow)$:USOLVE$_3$ (back-substitution sweep)
loop 300 $(K\uparrow)$:USOLVE$_3$ (periodicity sweep)
               USOLVE$_1$
               USOLVE$_2$
               DIV
               FFT2D
               PSOLVE (forward-elimination sweep)
loop 400 $(K\downarrow)$:PSOLVE (back-substitution sweep)
loop 500 $(K\uparrow)$:PSOLVE (periodicity sweep)
               FFT2D^{-1}
               GRAD

These loops are further detailed in Fig. 3. The following
description should be read in conjunction with that figure. The 100
and 200 loops are straightforward. In the 300 loop, the procedure
$USOLVE_3$ is completed. Once this has been accomplished on a given
plane it is possible to execute $USOLVE_1$ and $USOLVE_2$ on that plane
without moving the data. The next step in the algorithm is the
divergence computation. This is performed on a given plane after all
three steps of USOLVE are complete on the plane in question, the
plane above, and the plane below. Fortunately, our data structure
guarantees that all 3 of these planes are simultaneously present in
central memory.

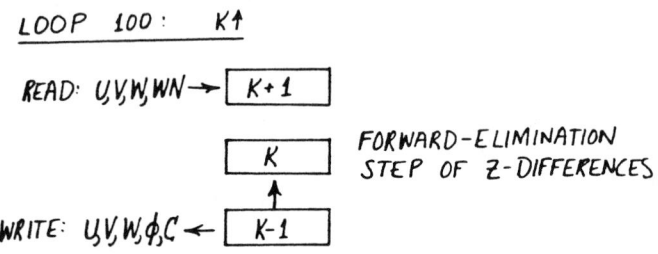

Figure 3a

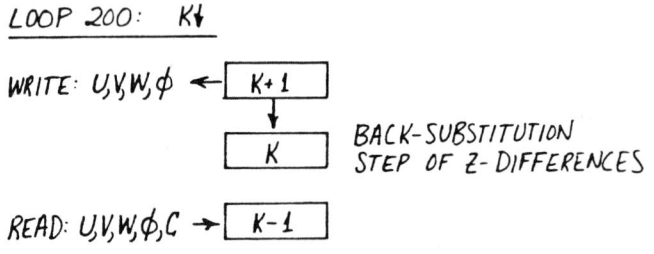

Figure 3b

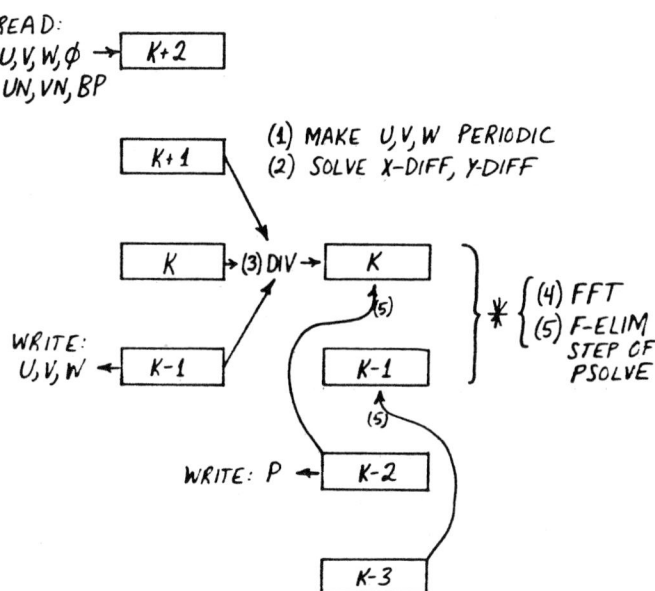

LOOP 300: K↑

READ:
U,V,W,φ → [K+2]
UN,VN,BP

[K+1]

(1) MAKE U,V,W PERIODIC
(2) SOLVE X-DIFF, Y-DIFF

[K] → (3) DIV → [K]

WRITE:
U,V,W ← [K-1]   [K-1]

⎱ ⎰ * ⎰ (4) FFT
     ⎱ (5) F-ELIM
          STEP OF
          PSOLVE

(5)

WRITE: P ← [K-2]

[K-3]

*Steps (4) and (5) are performed only when K is <u>odd</u>.

Figure 3c

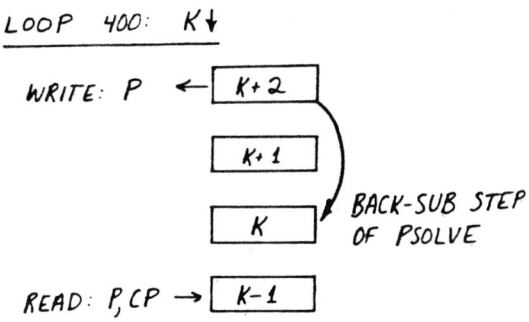

LOOP 400: K↓

WRITE: P ← [K+2]

[K+1]

[K]   BACK-SUB STEP
             OF PSOLVE

READ: P,CP → [K-1]

Figure 3d

125

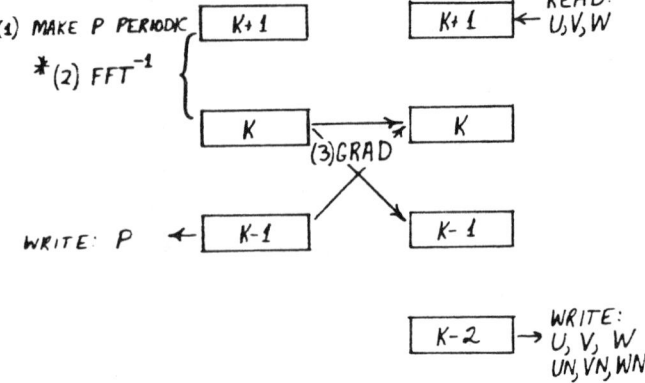

LOOP 500:   K ↑

READ: P, ∅P → [K+2]

(1) MAKE P PERIODIC [K+1]   [K+1] ← READ: U,V,W
＊(2) FFT⁻¹ {

                          [K]   [K]
                              (3)GRAD

WRITE: P ← [K-1]   [K-1]

                               [K-2] → WRITE: U, V, W UN, VN, WN

＊Step (2) is performed only when K is even.

Figure 3e

The remaining steps in the 300 loop operate on two planes at a time. Hence they are only executed for odd values of the loop index K. These steps are the computation of the two-dimensional Fourier transform and the forward elimination step of the procedure PSOLVE. The Fourier transform FFT2D is applied to planes (K-1,K), and PSOLVE then updates these planes using data from planes (K-3,K-2) which are still resident in central memory.

Overall, the 300 loop makes reference to 6 planes: (K-3) through (K+2). Despite this, the algorithm is consistent with our 4-plane central-memory data structure because only 4 planes are needed for any one variable: planes (K-1) through (K+2) for velocity data and planes (K-3) through K for the pressure data (see Fig. 3c).

The 400 loop is again straightforward: the only computation performed is the back-substitution sweep of PSOLVE. In the 500

126

loop, the PSOLVE operation is completed and the inverse Fourier transform is performed, two planes at a time. The most complicated part of the 500 loop is the computation of the gradient: $\underline{u}^{n+1} := \underline{u}^{n,3} - (\Delta t/\rho)\underline{G}p^{n+1}$. For the $\underline{e}_1$ and $\underline{e}_2$ components of $\underline{u}$, this computation involves no coupling between different planes. The vertical component of velocity, however, is influenced by the pressure on the plane above and on the plane below. To stay within the 4-plane data structure we compute these influences separately as shown in Fig. 3e. First, the velocity data on plane K are updated using the pressure on plane K-1 (the plane below). At the next pass through the loop, these same velocity data are in position K-1 (not that they have moved: K is bigger by 1), and they are updated again using the pressure on plane K (the plane above).

In summary, Subroutine FLUID has a pipeline structure. (Of course, this is a software pipeline that has nothing to do with any hardware feature of the machine.) The pipeline is circular and it has length 4. An important feature of this pipeline is that the data do not actually move. Instead the different functions (i/o and computation) work their way around the circle so that all functions are eventually performed on each plane of data.

Test problem: interacting plane waves

It is easy to generate exact solutions of the forced Navier-Stokes equations (Eqs. 1-2) for test purposes. We simply pick any (periodic) velocity and pressure $\underline{u}(\underline{x},t)$, $p(\underline{x},t)$ such that $\nabla \cdot \underline{u} = 0$, and then we substitute $(\underline{u},p)$ into Eq. (1) to determine the corresponding forcing function $\underline{F}$.

For $w = 1,2$, let

$$\tag{72} \underline{u}_w = \underline{a}_w \sin \theta_w$$

$$\tag{73} p_w = b_w \sin \theta_w$$

where

$$\tag{74} \theta_w = \underline{k}_w \cdot \underline{x} + \omega_w t + \phi_w$$

We insist that $\underline{k}_w$ be of the form

$$\tag{75} \underline{k}_w = \frac{2\pi}{L} \underline{K}_w$$

where $\underline{K}_w$ is a vector with integer coefficients. This ensures that $\underline{u}$, p will have the correct periodicity. Also, we insist that

$$\tag{76} \underline{k}_w \cdot \underline{a}_w = 0$$

This guarantees that $\nabla \cdot \underline{u}_w = 0$.

Then Eqs. (1-2) are satisfied if we set

$$\tag{77} \underline{u} = \sum_{w=1}^{2} \underline{u}_w$$

$$\tag{78} p = \sum_{w=1}^{2} p_w$$

$$\text{(79)} \quad \underline{F} = \sum_{w=1}^{2} (\rho \underline{a}_w \omega_w + b_w \underline{k}_w) \cos \theta_w + \mu |\underline{k}_w|^2 \underline{a}_w \sin \theta_w$$

$$+ \rho \sin \theta_1 \cos \theta_2 (\underline{a}_1 \cdot \underline{k}_2) \, \underline{a}_2$$

$$+ \rho \sin \theta_2 \cos \theta_1 (\underline{a}_2 \cdot \underline{k}_1) \, \underline{a}_1$$

To verify this formula for $\underline{F}$, we evaluate the different terms in Eq. (1). The linear terms can be evaluated separately for each w:

$$\text{(80)} \quad \rho \frac{\partial \underline{u}_w}{\partial t} = \rho \underline{a}_w \omega_w \cos \theta_w$$

$$\text{(81)} \quad \nabla p_w = b_w \underline{k}_w \cos \theta_w$$

$$\text{(82)} \quad -\mu \Delta \underline{u}_w = + \mu |\underline{k}_w|^2 \underline{a}_w \sin \theta_w$$

These expressions account for the terms inside the summation in Eq. (79). Next we consider the nonlinear terms. We have

$$\text{(83)} \quad \rho \underline{u} \cdot \nabla \underline{u} = \rho \sum_{w=1}^{2} \sum_{w'=1}^{2} \underline{u}_w \cdot \nabla \underline{u}_{w'}$$

$$= \rho \sum_{w=1}^{2} \sum_{w'=1}^{2} (\sin \theta_w)(\underline{a}_w \cdot \nabla \sin \theta_{w'}) \, \underline{a}_{w'}$$

$$= \rho \sum_{w=1}^{2} \sum_{w'=1}^{2} \sin \theta_w (\underline{a}_w \cdot \underline{k}_{w'})(\cos \theta_{w'}) \underline{a}_{w'}$$

Since $\underline{a}_w \cdot \underline{k}_w = 0$, the terms w = w' drop out and we are left only with the cross terms:

$$\text{(84)} \quad \rho \underline{u} \cdot \nabla \underline{u} = \rho(\sin \theta_1)(\cos \theta_2)(\underline{a}_1 \cdot \underline{k}_2) \underline{a}_2$$
$$+ \rho(\sin \theta_2)(\cos \theta_1)(\underline{a}_2 \cdot \underline{k}_1) \, \underline{a}_1$$

This completes the derivation of Eq. (79).

Incidentally, it is because $\underline{u}_w \cdot \nabla \underline{u}_w = 0$ that we need at least 2 plane waves to have a nontrivial test of a Navier–Stokes solver. For a single plane wave, the nonlinear terms vanish so a programming error involving the nonlinear terms might not show up. With two interacting plane waves, the nonlinear terms only vanish if $(\underline{a}_1 \cdot \underline{k}_2) = \underline{a}_2 \cdot \underline{k}_1 = 0$. Suppose $\underline{k}_1$ and $\underline{k}_2$ are independent. Since we also have $(\underline{a}_1 \cdot \underline{k}_1) = (\underline{a}_2 \cdot \underline{k}_2) = 0$, the only way that the nonlinear terms can vanish is if $\underline{a}_1$ and $\underline{a}_2$ are both perpendicular to the plane spanned by $\underline{k}_1$ and $\underline{k}_2$ (and hence parallel to each other). We avoid this case in the tests described below.

In the tests that follow, the domain is a periodic box with side $L = 1$ cm, and we use the plane-wave parameters listed in Table 1. The fluid density and viscosity are both 1 in cgs units: $\rho = 1$ gm/cm^3, $\nu = \mu/\rho = 1$ cm^2/sec. The initial velocity of the fluid is that of the exact solution evaluated at $t = 0$ (see Eqs. 72-77 and Table 1). For $t > 0$, we apply the force density given by Eq. (79), and we compute the solution up to time $t = 0.125$ sec.

Table 1.   Plane-wave Parameters*

	Wave # 1	Wave # 2	
a	(-2,1,2)	(2,-2,1)	cm/sec
b	1.0	1.0	dynes/cm^2
K	(2,2,1)	(1,2,2)	
ω	0.1	0.1	sec^{-1}
∅	0.0	0.0	

*See Eqs. (72-76)

We report here on two types of results: accuracy and computer time. Of course, the accuracy of the results is only a consequence of the accuracy of Chorin's method; it has nothing to do with the input/output techniques that are the main subject of this paper. Nevertheless, it is worthwhile to measure the accuracy as an overall check on the correctness of the program and also to get a sense of the range of parameters for which the method gives reasonable

130

answers.

The formal order of accuracy of Chorin's method is $O(\Delta t) + O((\Delta x)^2)$. In our tests, we set $\nu \, \Delta t = (\Delta x)^2$. With this setup, we expect that the error will be $O((\Delta x)^2)$. To check this, we solve the problem outlined above with NG = 16,32,64. In Fig. 4, we plot the logarithm of the L_2 norm of the error against the logarithm of NG. This is done for the pressure and for the x_1 component of the velocity (the other two components of velocity yield very similar errors). For large NG, these plots should be linear with slope -2, so a reference line with this slope is plotted for comparison. The L_2 norm of the velocity error shows clearly the expected behavior of a second-order-accurate method. In the pressure, the errors are much larger, and the second-order behavior is somewhat less precise, but it does appear from the plot that the error in the pressure is approaching that of a second-order method. In any case, the pressure in this method is just an auxiliary device that is used in the computation of a correct velocity field. From this standpoint, the method should really be judged by the velocity error alone.

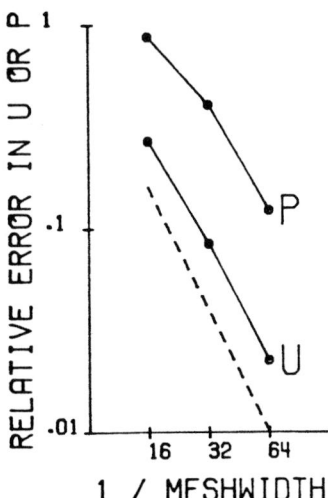

Figure 4

We now study the influence of the Reynolds number on the relative accuracy of the computed solution. To do this, we start from the exact solution used above (see Table 1), and we multiply all velocity components by the parameter USCALE. Since $R = L\|\underline{u}\|_{max}/\nu$, this makes R proportional to USCALE. In our particular case, $L = 1$ cm, $\nu = 1$ cm^2/sec., and $\|\underline{u}\|_{max} = (4$ cm/sec.$) *$ USCALE, so $R = 4 *$ USCALE. According to Inequality (33), the method is only expected to work for $R < 2(NG)$. Thus, with $NG = 64$, we expect breakdown at USCALE $= 2 * 64/4 = 32$. This expectation is confirmed in Fig. 5, which is a log-log plot of relative error vs. USCALE.

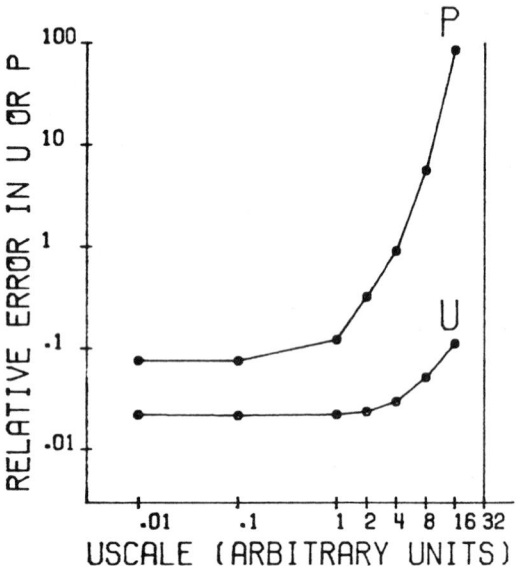

Figure 5

We now turn to a consideration of timing statistics. All of the results cited below are for the case $NG = 64$, and all statistics are on a per-time-step basis. The timing statistics for the Cray 1a given in Table 2. We used FLOWTRACE to obtain a breakdown of CPU time by subroutine, and then we lumped together the results for routines that are functionally related. As one might expect from the structure

132

of the algorithm, the bulk of the CPU time is spent solving tridiagonal systems (43%) and computing forward and backward Fourier transforms (36%). [In connection with the solution of the tridiagonal systems, it is remarkable that the X and Y tridiagonal systems are solved in the same amount of CPU time. This shows that we have successfully avoided memory-bank conflicts, as claimed above.] The divergence and gradient computations combined take only a small amount (8%) of the total CPU time. The rest (13%) is the CPU time used for i/o processing. Note that the last figure does not correspond to the total time required for i/o (see below); it merely reflects the percentage of time that the *central processor* is busy with operations connected with i/o. The bulk of the i/o is handled by separate processors and can proceed in parallel with computation.

Table 2. Timing Statistics, Cray 1a[1]

64^3 lattice: one time step

CPU Time

    Tridiagonal systems:

X:	0.20 sec	
Y:	0.20 sec	
Z:	0.27 sec[2]	
FFT:	0.57 sec[3]	
DIV:	0.04 sec	
GRAD:	0.08 sec	
I/O:	0.21 sec[4]	
Total CPU:	1.57 sec	

I/O Wait Time

    Synchronous: 2 min

    Asynchronous: 1 min

[1]At Research Equipment Inc., University of Minnesota.

[2]Includes both USOLVE$_3$ and PSOLVE.

[3]Includes both forward and inverse transform on all planes.

[4]Includes only CPU time associated with i/o. See also i/o wait time.

The i/o wait time in Table 2 is the amount of time that the job is resident in central memory waiting for i/o. During this time, the central processor is free to work on other jobs, provided, of course, that such jobs fit into central memory along with the job in question. Note that the i/o wait time (per time step) is given in minutes while the CPU time is given in seconds. With synchronous i/o, the i/o wait time is 2 minutes; with asynchronous i/o it is cut neatly in half to 1 minute. Note that this saving is far too great to be explained by the overlap of i/o with computation, as there just isn't enough computation available to do the job (1.57 seconds as computed with 2 minutes). Thus, essentially all of the savings comes from the overlap of i/o with itself.

From the structure of our algorithm, one might think that the savings associated with asynchronous i/o would be substantially more than a factor of 2. A factor of 2 could be explained merely by the overlap of input with output on any given file. (Recall that such overlap was made possible by the dual representation of each disk file.) We might expect a further savings from the overlap of one file's i/o with that of another.

There are at least two possible explanations for the failure to obtain as much improvement as expected by the use of asynchronous i/o. The first explanation concerns the number of available channels for communication between disk and central memory. Even if the hardware has many such channels, the operating system may limit the number available to any single user. (We thank S. Orszag for pointing out this possibility.) If, for example, there were only one channel available for input and another for output, then one would expect a maximum speedup of 2 from the concurrent use of both channels.

Another possible explanation of limited speedup is the startup time associated with each read or write operation. If these startup times are significant and if they cannot be overlapped with each other, then the speedup associated with asynchronous i/o will be less than expected. This effect could be measured by comparing the speedup for different values of NG. We have not yet performed such studies, however.

Statistics for the exact same Fortran code run on the Cray X-MP are compiled in Table 3. On the Cray X-MP, however, we can compare the disk performance with that of an SSD. The performance is measured both in terms of cost and also in terms of time. In the time results, we do not have a figure for i/o wait time, so we use "wall" time instead. Since wall time measures the total elapsed time (during one time step) of the job, it can only be an *over*-estimate of the i/o wait time.

Table 3. Comparison of Disk and SSD Performance

Cray X-MP: one Processor[1]
64^3 lattice: one time step

Cost in "CCU"s[2]

	DISK	SSD	SSD/DISK
CPU	84	85	1.01
I/O	863	89	0.10
TOTAL	947	174	0.18
I/O cost / TOTAL cost	91%	51%	

Time in seconds

	DISK	SSD	SSD/DISK
CPU	1.3	1.3	1.00
"WALL"[3]	30.4	7.6	0.25

[1] At Boeing Computer Services.
[2] Computational Cost Unit.
[3] Physical time elapsed per time step of the job.

136

The results of the comparison in Table 3 are very dramatic. In this application, the SSD is 10 times cheaper to use than a conventional disk, and it reduces the overall elapsed time of the job by at least a factor of 4. Another point worth noticing in Table 3 is that, with the help of the SSD, we are able to achieve a balanced distribution of cost between i/o and computation. This is a reasonable goal, since it implies that only a limited reduction in overall cost could be achieved by a further reduction of the i/o charges.

Recently, we had the opportunity to perform further timing tests of the code on the Cray X-MP/SSD at the Cray Software Training Center, Mendota Heights, Minnesota. These tests were performed in dedicated mode, so they are not corrupted by memory-bank conflicts with other users. Moreover, we used the facility FLOPTRACE, an extension of FLOWTRACE, which uses the hardware-monitoring features of the Cray X-MP to obtain megaflop rates on an individual-subroutine basis. (The use of FLOPTRACE substantially increases the wall time, however, so wall-time results were obtained with FLOPTRACE turned off.)

These timing tests were performed after certain improvements had been made in the FFT segment of the code. These include: (i) the use of scalar temporaries together with the compiler option KILLTEMP to avoid unnecessary central-memory references, (ii) the relegation of bit-reversal to initialization only, and (iii) the use of tables of sines and cosines. Despite these changes, vectorization is still achieved in the manner described above.

Another important change is that these timing tests are for the case NG = 128 instead of NG = 64. This doubles the vector length, and it improves the efficiency of i/o since each i/o request now involves 4 times as many words.

The timing results are summarized in Table 4. Note, first, that the computational machinery of Subroutine FLUID (i.e., the tridiagonal systems, Fourier transforms, divergence, and gradient) runs at an average rate of 116 Mflops/sec. This should be compared with the theoretical single-processor maximum of 200 Mflops/sec., which would require all functional units busy at all times. The second point to note is that the overhead associated with i/o is rather modest. This overhead has two parts: the time spent by the CPU in the i/o routines (which is 0.58 sec.) and the time when the CPU is idle. The latter figure is the difference between the WALL time and the total CPU time. It is 4.24 sec. with synchronous i/o but only 2.36 sec. with asynchronous i/o. In the case of asynchronous i/o, the total overhead is 2.94 sec., which is 33% of the wall time. The bottom line is the effective megaflop rate, which is the number of megaflops divided by the wall time. With asynchronous i/o, this rate is 77 Mflops/sec.

Table 4. Timing statistics, Cray X-MP/SSD[1]

One processor: dedicated machine
128^3 lattice: one time step

CPU time and Megaflop rate

Tridiagonal systems:

X:	0.59 sec	120 Mflops/sec
Y:	0.57 sec	125 Mflops/sec
Z:	0.77 sec[2]	116 Mflops/sec
FFT:		
direct:	0.61 sec[3]	124 Mflops/sec
inverse:	0.66 sec[3]	115 Mflops/sec
DIV:	0.15 sec	82 Mflops/sec
GRAD:	0.19 sec	88 Mflops/sec
FLUID:	3.54 sec	116 Mflops/sec
MAIN:	2.32 sec[4]	114 Mflops/sec
I/O:	0.58 sec[5]	0 Mflops/sec
Total CPU:	6.44 sec	105 Mflops/sec

WALL time[6] and Effective Computation Rate[7]

Synchronous:	10.68 sec	63 Mflops/sec
Asynchronous:	8.80 sec	77 Mflops/sec

[1]At the Cray Software Training Center, Mendota Heights, Minnesota.

[2]Includes both USOLVE$_3$ and PSOLVE.

[3]Includes transform on all planes of the lattice.

[4]Includes computation and application of external force to the fluid. Also includes comparison of computed results with exact solution.

[5]Includes only CPU time of the i/o routines. Additional overhead is the difference between the WALL time and the total CPU.

[6]Physical time elapsed during the computation.

[7]Total Mflops divided by WALL time.

Summary and Conclusions

The subject of this paper is the out-of-core, vectorized implementation of Chorin's projection method for the solution of the three-dimensional, time-dependent, incompressible Navier-Stokes equations in a periodic box. The implementation discussed here is particularly suitable for a computer such as the Cray X-MP equipped with an SSD. This computer has a modest amount of fast, random-access central memory, and the SSD provides a large, slower, sequential-access memory that is still much faster than a conventional disc. Asynchronous i/o makes it possible for information transfer to proceed concurrently over two i/o channels between the SSD and the central memory and also for such i/o to proceed in parallel with computation.

In our implementation of Chorin's method, all three-dimensional (N^3) arrays are stored in the SSD. Each such SSD array has a corresponding central-memory array with a short (independent of N) third dimension, so the central-memory requirements of the implementation are only proportional to N^2, not N^3. Central memory is organized as a circular buffer: The index of the short third dimension cycles through central memory as the plane index sweeps through all planes of the computational lattice. This avoids unnecessary data movement. All inner loops are successfully vectorized by the Cray Fortran compiler and the length of these loops is an integer multiple of 64, so the vector registers are fully utilized.

The reduction in central-memory requirements from $O(N^3)$ to $O(N^2)$ has a dramatic effect on the size of the problem that can be solved in any given computer. Moreover, since central-processor charges are often proportional to central-memory utilization, the reduction in central-memory requirement may also have a profound effect on the cost of the computation even in the situation where the problem could be run entirely in-core.

One might think that these advantages would be offset by the overhead associated with i/o. With the technqiues described in this paper, however, we have kept such overhead within tolerable limits: In the Cray X-MP/SSD experiment with NG = 64 reported here, the physical elapsed time for the job is about 6 times the central-processor time, and the i/o charges are roughly equal to the central processor charges. With NG = 128, the i/o efficiency is dramatically improved, and the physical elapsed time is only 1.5 times the central-processor time.

The work described here has implications for the design of supercomputers. At least in this application, it shows that an architecture such as that of the Cray X-MP/SSD can be effectively utilized to solve problems much larger than those that fit into central memory. It remains to be seen, however, whether this is the best way to solve such problems or whether a large-central-memory machine such as the Cray 2 is preferable.

Acknowledgment

It is a pleasure to acknowledge the key contribution of Olof Windlund, whose out-of-core Poisson solver is the direct ancestor of the Navier-Stokes solver reported here. Chris Hsiung of Cray Research was extremely helpful with asynchronous i/o. Our READIN and WRTOUT routines (see Appendix) are patterned after a more general i/o interface of Hsiung's.

Our development work for this paper was performed on the Cray 1 located at Research Equipment, Inc. (now called the Minnesota Supercomputer Center) and would not have been possible without the anonymous help of their HELP-line consultants. We are also indebted to Boeing Computer Services for time on their Cray X-MP/SSD; this time was used to obtain the results shown in Table 3. The results shown in Table 4 were obtained at a workshop held at Cray Research and organized by Philippe de Forcrand. We are indebted to Cray both for the use of their machine and for their expert help in improving the code.

This work was supported by NSF Grant DMS-8312229 (Multiphase Flow) and by N.I.H. Grant HL17859 (Computational Methods in Cardiac Fluid Dynamics).

142

References

1. Edwards, M., Hsiung, C.C., Kosloff, D., and Reshelf, M., "Seismic 3-D Fourier modeling on the Cray X-MP." Submitted to J. Supercomputing. (See also Cray Channels, Spring 1986, p. 2-5.)

2. Peskin, C.S., "Numerical analysis of blood flow in the heart." J. Comput. Phys. 25: 220-252, 1977.

3. Peskin, C.S., and McQueen, D.M., "Modeling prosthetic heart valves for numerical analysis of blood flow in the heart." J. Comput. Phys. 37: 113-132, 1980.

4. Chorin, A.J., "Numerical solution of the Navier-Stokes equations." Math. Comp. 22: 745-762, 1968.

5. Chorin, A.J., "On the convergence of discrete approximations to the Navier-Stokes equations." Math. Comp. 23: 341-353, 1969.

6. Fischer, D., Golub, G., Hald, O., Leiva, C., and Widlund, O., "On Fourier-Toeplitz methods for separable elliptic problems." Math. Comp 28: 349-368, 1974.

7. Widlund, O., unpublished communication.

8. Cooley, J.W., Lewis, P.A.W., and Welch, P.D., "The fast Fourier transform and its applications." IEEE Transactions E-12: 27-34, 1969.

9. Dahlquist, G., and Björk, Å, Numerical Methods (Trans: Anderson, N.), Prentice-Hall, Englewood Cliffs, NJ, 1974.

10. O'Leary, D.P., and Widlund, O., "Capacitance matrix methods for the Helmholtz equation on general three dimensional regions." Math. Comp. 33: 849-879, 1979.

APPENDIX

The following is a complete listing of our i/o interface. The routines called by the user are READIN, WRTOUT, WAITR, WAITW, and SWITCH; their use has been explained in the main body of the paper.

READIN and WRTOUT simply compute the first address of the i/o transfer. They pass this information on to RDIN and WOUT, respectively.

RDIN and WOUT each perform the following operations:

(i)   Look up FORTRAN unit number corresponding to the symbolic unit number.

(ii)   Wait if the FORTRAN unit is busy.

(iii)   Check that the record number is in range. (If not, RETURN.)

(iv)   Set file position to beginning of i/o.

(v)   Initiate i/o.

(vi)   If a flag has been set for synchronous i/o, wait for i/o to finish. (If not, RETURN and let computation proceed in parallel with i/o.)

The routines WAITR and WAITW look up the appropriate FORTRAN unit number (for reading or writing, respectively) corresponding to the symbolic unit number. They then call WAIT, which waits until i/o is complete on the FORTRAN unit in question.

SWITCH interchanges the read and write entries corresponding to a particular symbolic unit on the table of read and write unit numbers, URW.

The listing follows:

```
        SUBROUTINE READIN(U,K,A,KX)
        INTEGER U
        PARAMETER(L2NG=6,NG=2**L2NG,NB=NG+1)
        PARAMETER(NSIZE=NG*(NG+2),NW=512*(1+NSIZE/512),NGY=1+NW/(NG+2))
        DIMENSION A(0:NB,1:NGY,0:3)
        CALL RDIN(U,K,A(0,1,KX))
        RETURN
        END

        SUBROUTINE WRTOUT(U,K,A,KX)
        INTEGER U
        PARAMETER(L2NG=6,NG=2**L2NG,NB=NG+1)
        PARAMETER(NSIZE=NG*(NG+2),NW=512*(1+NSIZE/512),NGY=1+NW/(NG+2))
        DIMENSION A(0:NB,1:NGY,0:3)
        CALL WOUT(U,K,A(0,1,KX))
        RETURN
        END

        SUBROUTINE RDIN(USYM,K,A)
C  THIS ROUTINE READS RECORD K FROM UNIT U INTO THE ARRAY A
C  IT IS ASSUMED THAT ALL RECORDS HAVE THE SAME LENGTH,
C  WHICH IS NW WORDS (SEE PARAMETER STATEMENTS)
        PARAMETER(L2NG=6,NG=2**L2NG,NSIZE=NG*(NG+2))
        PARAMETER(NW=512*(1+NSIZE/512))
        PARAMETER(NGM1=NG-1)
        COMMON/SYNCSW/SYNCON
        COMMON/RW/URW(2,12)
        LOGICAL SYNCON
        INTEGER U,URW,USYM
        DIMENSION A(NW)
C  USYM=SYMBOLIC UNIT NUMBER
        U=URW(1,USYM)
        CALL WAIT   (U)
        IF(K.LT.0)RETURN
        IF(K.GT.(NG-1))RETURN
        CALL SETPOS(U,K*NW)
        BUFFERIN(U,0)(A(1),A(NW))
        IF(SYNCON)CALL WAIT   (U)
        RETURN
        END

        SUBROUTINE WOUT(USYM,K,A)
C  THIS ROUTINE WRITES THE ARRAY A ONTO RECORD K OF UNIT U
C  IT IS ASSUMED THAT ALL RECORDS HAVE THE SAME LENGTH,
C  WHICH IS NW WORDS (SEE PARAMETER STATEMENTS)
        PARAMETER(L2NG=6,NG=2**L2NG,NSIZE=NG*(NG+2))
        PARAMETER(NW=512*(1+NSIZE/512))
        PARAMETER(NGM1=NG-1)
        COMMON/SYNCSW/SYNCON
        COMMON/RW/URW(2,12)
        LOGICAL SYNCON
        INTEGER U,URW,USYM
        DIMENSION A(NW)
C  USYM IS SYMBOLIC UNIT NUMBER
        U=URW(2,USYM)
        CALL WAIT   (U)
        IF(K.LT.0)RETURN
        IF(K.GT.(NG-1))RETURN
        CALL SETPOS(U,K*NW)
        BUFFEROUT(U,0)(A(1),A(NW))
        IF(SYNCON)CALL WAIT   (U)
        RETURN
        END
```

```
      SUBROUTINE WAITR(USYM)
      INTEGER U,URW,USYM
      COMMON/RW/URW(2,12)
      U=URW(1,USYM)
      CALL WAIT(U)
      RETURN
      END

      SUBROUTINE WAITW(USYM)
      INTEGER U,URW,USYM
      COMMON/RW/URW(2,12)
      U=URW(2,USYM)
      CALL WAIT(U)
      RETURN
      END

      SUBROUTINE WAIT(U)
C  THIS ROUTINE HOLDS UP EXECUTION UNTIL THE LAST READ OR WRITE
C  OPERATION ON UNIT U IS COMPLETE.   (ALSO, ANY ERRORS IN THAT
C  READ OR WRITE WILL STOP THE PROGRAM.)
      INTEGER U
      USTAT=UNIT(U)
      IF(USTAT .LT. 0.)RETURN
      WRITE(6,*)'I/O ERROR ON UNIT ',U,' ERROR STATUS CODE ',USTAT
      STOP
      END

      SUBROUTINE SWITCH(U)
C  THIS ROUTINE SWITCHES THE READ AND WRITE UNITS
C  OF THE FILE WITH SYMBOLIC UNIT NUMBER U.
C     URW(1,U)=READ UNIT NUMBER
C     URW(2,U)=WRITE UNIT NUMBER
C  SEE ROUTINES READIN AND WRTOUT.
      INTEGER U,URW
      COMMON/RW/URW(2,12)
      IT=URW(1,U)
      URW(1,U)=URW(2,U)
      URW(2,U)=IT
      RETURN
      END
```

ON THE NONLINEARITY OF
MODERN SHOCK-CAPTURING SCHEMES

Ami Harten

School of Mathematical Sciences, Tel-Aviv University

and

Department of Mathematics, UCLA

Dedicated to Peter Lax on his 60th birthday

ABSTRACT

In this paper we review the development of shock-capturing methods, paying special attention to the increasing nonlinearity in the design of the numerical schemes. We study the nature of the nonlinearity and examine its relation to upwind differencing. This nonlinearity is essential in the sense that analysis by local linearization is not justified and may even lead to incorrect conclusions; examples to demonstrate this point are given.

Research supported in part by NSF Grant DMS-8120790.

1.   Introduction

In this paper, we describe and analyze numerical techniques that are designed to approximate weak solutions of hyperbolic systems of conservation laws in several space dimensions.   For sake of exposition, we shall describe these methods as they apply to the pure initial value problems (IVP) for a one–dimensional scalar conservation law

(1.1) $$u_t + f(u)_x = 0, \qquad u(x,0) = u_0(x).$$

To further simplify our presentation, we assume that the flux $f(u)$ is a convex function, i.e., $f''(u) > 0$ and that the initial data $u_0(x)$ are piecewise smooth functions which are either periodic or of compact support.   Under these assumptions, no matter how smooth u_0 is, the solution $u(x,t)$ of the IVP (1.1) becomes discontinuous at some finite time $t = t_c$.   In order to extend the solution for $t > t_c$, we introduce the notion of weak solutions, which satisfy

(1.2a) $$\frac{d}{dt} \int_a^b u \; dx + f(u(b,t)) - f(u(a,t)) = 0$$

for all $b \geqslant a$ and $t \geqslant 0$.   Relation (1.2a) implies that $u(x,t)$ satisfies the PDE in (1.1) wherever it is smooth, and the Rankine–Hugoniot jump relation

(1.2b) $$f(u(y + 0,t)) - f(u(y - 0,t)) = [u(y + 0,t) - u(y - 0,t)] \frac{dy}{dt}$$

across curves $x = y(t)$ of discontinuity.

It is well known that weak solutions are not uniquely determined by their initial data.   To overcome this difficulty, we consider the IVP (1.1) to be the vanishing viscosity limit $\epsilon \downarrow 0$ of the parabolic problem

(1.3a) $$(u^\epsilon)_t + f(u^\epsilon)_x = \epsilon(u^\epsilon)_{xx}, \qquad u^\epsilon(x,0) = u_0(x),$$

and identify the unique "physically relevant" weak solution of (1.1) by

(1.3b)
$$u = \lim_{\epsilon \downarrow 0} u^{\epsilon}.$$

The limit solution (1.3) can be characterized by an inequality that the values $u_L = u(y - 0,t)$, $u_R = u(y + 0,t)$ and $s = dy/dt$ have to satisfy; this inequality is called entropy condition; admissible discontinuities are called shocks. When $f(u)$ is convex, this inequality is equivalent to Lax's shock condition

(1.4)
$$a(u_L) > s > a(u_R)$$

where $a(u) = f'(u)$ is the characteristic speed (see [20] for more details).

We turn now to describe finite difference approximations for the numerical solution of the IVP (1.1). Let v_j^n denote the numerical approximation to $u(x_j,t_n)$ where $x_j = jh$, $t_n = n\tau$; let $v_h(x,t)$ be a globally defined numerical approximation associated with the discrete values $\{v_j^n\}$, $-\infty < j < \infty$, $n \geq 0$.

The classical approach to the design of numerical methods for partial differential equations is to obtain a solvable set of equations for $\{v_j^n\}$ by replacing derivatives in the PDE by appropriate discrete approximations. Therefore, there is a conceptual difficulty in applying classical methods to compute solutions which may become discontinuous. Lax and Wendroff [21] overcame this difficulty by considering numerical approximations to the *weak formulation* (1.2a) rather than to the PDE (1.1). For this purpose, they have introduced the notion of scheme in conservation form:

(1.5a)
$$v_j^{n+1} = v_j^n - \lambda(\bar{f}_{j+1/2} - \bar{f}_{j-1/2}) \equiv (E_h \cdot v^n)_j;$$

here $\lambda = \tau/h$ and $\bar{f}_{i+1/2}$ denotes

(1.5b)
$$\bar{f}_{i+1/2} = f(v_{i-k+1}^n, \dots, v_{i+k}^n);$$

149

$\bar{f}(w_1,...,w_{2k})$ is a numerical flux function which is consistent with the flux $f(u)$, in the sense that

$$(1.5c) \qquad\qquad \bar{f}(u,u,...,u) = f(u);$$

E_h denotes the numerical solution operator. Lax and Wendroff proved that if the numerical approximation converges boundedly almost everywhere to some function u, then u is a weak solution of (1.1), i.e., it satisfies the weak formulation (1.2a). Consequently discontinuities in the limit solution automatically satisfy the Rankine-Hugoniot relation (1.2b). We refer to this methodology as shock-capturing (a phrase coined by H. Lomax).

In the following, we list the numerical flux function of various 3-point schemes ($k = 1$ in (1.5b)):

(i)   The Lax-Friedrichs scheme [19]

$$(1.6) \qquad\qquad \bar{f}(w_1,w_2) = \frac{1}{2} \left[f(w_1) + f(w_2) - \frac{1}{\lambda} (w_2 - w_1) \right]$$

(ii)   Godunov's scheme [5]

$$(1.7a) \qquad\qquad \bar{f}(w_1,w_2) = f(V(0;w_1,w_2));$$

here $V(x/t;w_1,w_2)$ denotes the self-similar solution of the IVP (1.1) with the initial data

$$(1.7b) \qquad\qquad u_0(x) = \begin{cases} w_1 & x < 0 \\ w_2 & x > 0 \end{cases}.$$

(iii)   The Cole-Murman scheme [26]:

$$(1.8a) \qquad \bar{f}(w_1,w_2) = \frac{1}{2} \left[f(w_1) + f(w_2) - |\bar{a}(w_1,w_2)| (w_2 - w_1) \right]$$

where

150

$$(1.8b) \qquad \bar{a}(w_1, w_2) = \begin{cases} \dfrac{f(w_2) - f(w_1)}{w_2 - w_1} & \text{if } w_1 \neq w_2 \\ a(w_1) & \text{if } w_1 = w_2 \end{cases}.$$

(iv)   The Lax-Wendroff scheme [21]:

$$(1.9) \quad \bar{f}(w_1, w_2) = \frac{1}{2} \{ f(w_1) + f(w_2) - \lambda a \frac{(w_1 + w_2)}{2} [f(w_2) - f(w_1)] \}.$$

(v)   MacCormack's scheme [24]:

$$(1.10) \qquad \bar{f}(w_1, w_2) = \frac{1}{2} \{ f(w_2) + f(w_1 - \lambda [f(w_2) - f(w_1)]) \}.$$

Let E(t) denote the evolution operator of the exact solution of (1.1) and let E_h denote the numerical solution operator defined by the RHS of (1.5a).   We say that the numerical scheme is r-th order accurate (in a pointwise sense) if its local truncation error satisfies

$$(1.11) \qquad\qquad E(\tau) \cdot u - E_h \cdot u = O(h^{r+1})$$

for all sufficiently smooth u; here $\tau = O(h)$.   If $r > 0$, we say that the scheme is consistent.

The scheme of Lax-Friedrichs (1.6), Godunov (1.7) and Cole-Murman (1.8) are first order accurate; the schemes of Lax-Wendroff (1.9) and MacCormack are second order accurate.

We remark that the Lax-Wendroff theorem states that if the scheme is convergent, then the limit solution satisfies the weak formulation (1.2b); however, it need not be the entropy solution of the problem (see [11]).   It is easy to see that the schemes of Cole-Murman (1.8), Lax-Wendroff (1.9) and MacCormack (1.10) admit a stationary "expansion shock" (i.e., $f(u_L) = f(u_R)$ with $a(u_L) < a(u_R)$) as a steady solution.   This problem can be easily rectified by adding sufficient numerical dissipation to the scheme (see [25] and [10]).

The cardinal problem that is yet to be resolved is the question of convergence of the numerical approximation.

2. Linear Stability and Computation of Weak Solutions

Let us consider the constant coefficient case $f(u) = au$, $a = $ const. in (1.1), i.e.,

(2.1a)
$$u_t + au_x = 0, \qquad u(x,0) = u_0(x),$$

the solution to which is

(2.1b)
$$u(x,t) = u_0(x - at).$$

In this case, all the schemes mentioned in the previous section, (1.6)–(1.10), take the form

(2.2)
$$v_j^{n+1} = \sum_{\ell=-k}^{k} C_\ell v_{j-\ell}^n \equiv (E_h \cdot v^n)_j,$$

where C_ℓ are constants independent of j (C_ℓ are polynomial functions of the CFL number $\nu = \lambda a$). We note that in the constant coefficient case Godunov's scheme is identical to that of Cole-Murman; the MacCormack scheme is identical to that of Lax-Wendroff. Since the numerical solution operator E_h of these schemes in the constant coefficient case becomes a linear operator, we shall refer to these schemes as essentially linear or just "linear" schemes.

Next we briefly review the convergence theory of linear schemes; we refer the reader to [29] for a detailed analysis.

We say that the numerical scheme is stable if

(2.3a)
$$\|(E_h)^n\| \leq C \qquad \text{for} \qquad 0 \leq n\tau \leq T, \ \tau = 0(h).$$

The constant coefficient scheme (2.2) is stable if and only if it satisfies von Neumann's condition:

(2.3b)
$$\left| \sum_{\ell=-k}^{k} C_\ell \, e^{-i\ell\xi} \right| \leq 1 \qquad \text{for all} \qquad 0 \leq \xi \leq \pi.$$

It is easy to verify that all the 3-point schemes (1.6)–(1.10) satisfy condition (2.3b) under the Courant-Friedrichs-Lewy (CFL)

152

restriction

(2.4) $$|\nu| = |\lambda a| \leqslant 1,$$

and thus are linearly stable. The notion of stability (2.3a) is related to convergence through Lax's equivalence theorem, which states that a consistent linear scheme is convergent if and only if it is stable.

The accumulation of error in a computation with a linearly stable scheme (2.2) is linear, in the sense that if the local truncation error (1.11) is $O(h^{r+1})$, then after performing $N = T/\tau = O(h^{-1})$ time-steps, the error is $O(h^r)$, i.e.,

(2.5) $$u(x_j, N\tau) - v_j^N = O(h^r).$$

An immense body of work has been done to find out whether stability of the constant coefficient scheme with respect to all "frozen coefficients" associated with the problem, implies convergence in the variable coefficient case and in the nonlinear case.

In the variable coefficient case, where the numerical solution operator is linear and Lax's equivalence theorem holds, it comes out that the stability of the variable coefficient scheme depends strongly on the dissipativity of the constant coefficient one, i.e., on the particular way it damps the high-frequency components in the Fourier representation of the numerical solution.

In the nonlinear case, under assumptions of sufficient smoothness of the PDE, its solution and the functional definition of the numerical scheme, Strang proved that linear stability of the first variation of the scheme implies its convergence; we refer the reader to [29] for more details.

Although there is no rigorous theory to support the supposition that linearly stable schemes should converge in the case of discontinuous solutions of nonlinear problems, we find in practice that this is true in many (although not all) instances; when such a scheme fails to converge, we refer to this case as "nonlinear instability". The occurence of a nonlinear instability is usually associated with insufficient numerical dissipation which triggers exponential growth of the high-frequency components of the numerical solution.

Next we present two shock-tube calculations by the scheme (1.5) with the numerical flux

(2.6) $\bar{f}(w_1,w_2) = \frac{1}{2} \{f(w_2) + f(w_1 - \lambda [f(w_2) - f(w_1)]) - \frac{\theta}{4} (w_2 - w_1)\}.$

The shock-tube problem is modelled by a Riemann IVP for the one-dimensional Euler equations of compressible gas:

(2.7a) $u_t + f(u)_x = 0, \qquad u(x,0) = \begin{cases} u_L & x < 0 \\ u_R & x < 0 \end{cases}$

where

(2.7b) $u = (\rho, q\rho, E)^T, \qquad f(u) = qu + (0, p, qp)^T$

with

(2.7c) $p = (\gamma - 1)(E - \frac{1}{2} \rho q^2).$

Here $\rho, q, p,$ and E are the density, velocity, pressure, and total energy, respectively. In these calculations, $\gamma = 1.4$ and

(2.7d) $u_L = (0.445, 0.3111, 8.928), \qquad u_R = (0.5, 0., 1.4275).$

The exact solution to this Riemann problem consists of a shock propagating to the right followed by a contact discontinuity and a left-propagating rarefaction wave; it is shown by a continuous line in Figures 1 and 2. The numerical solution of (2.6) is shown in Figures 1 and 2 by circles.

Figure 1 shows the results of the second-order accurate MacCormack scheme, i.e., $\theta = 0$ in (2.6). Observe the large spurious oscillations at the shock and at the contact discontinuity--this is a Gibbs-like phenomenon. Note that although the rarefaction wave is computed rather accurately, there are some spurious oscillations at its right-endpoint due to the discontinuity in the first derivative there.

Figure 2 shows the results of the first-order accurate scheme

(2.6) with $\theta = 1$. Observe that now the numerical solution is oscillation-free. However, both the shock and the contact discontinuity are now smeared much more than the corresponding ones in Figure 1. Note the excessive rounding of the corners at the endpoints of the rarefaction wave.

It is important to understand that the Gibbs-phenomenon by itself is not an instability; this is self-evident when we consider the constant coefficient problem (2.1) with discontinuous initial data u_0. However, in compressible gas calculations, where both density and pressure are restricted to have non-negative values, the Gibbs phenomenon may cause the numerical solution to get out of the physical domain. Attempting to replace negative values of density and pressure by positive ones makes the scheme nonconservative and may result in an exponential growth of the solution.

The comparison between Figure 1 $(\theta = 0)$ and Figure 2 $(\theta = 1)$ shows that the Gibbs phenomenon in the second-order accurate scheme can be controlled by the addition of a numerical viscosity term. To do so without losing the second-order accuracy, Lax and Wendroff [21] suggested to take in (2.6) $\theta = \theta(w_1, w_2)$ of the form

$$(2.8) \qquad \theta(w_1, w_2) = x\,|\,a(w_2) - a(w_1)\,|\,;$$

here $a = f'(u)$ and x is a dimensionless constant; observe that $\theta \equiv 0$ in the constant coefficient case.

Numerical experiments showed that as x increases the size of the spurious osciilations decreases, but at the cost of increased smearing of the discontinuity. Furthermore, when x is fixed, the size of the spurious oscillations increases with the strength of the shock. These observations indicate that the numerical viscosity term (2.8) does not have an appropriate functional dependence on the parameters that control the Gibbs phenomenon. Consequently, the choice of a suitable value of x is problem dependent, and the practical use of the numerical scheme requires several preliminary runs to "tune parameters".

Ideally, we would like to have high-order accurate schemes that are capable of propagating a shock wave without having any spurious oscillations. In the scalar case, this can be accomplished by designing schemes to be monotonicity preserving, i.e., to satisfy

(2.9)   v monotone $\Rightarrow$ E$_h$ · v monotone.

Godunov [5] has considered this avenue of design in the constant coefficient case (2.1) and showed that monotonicity preserving *linear* schemes (2.2) are necessarily only first order accurate. For some time this result has been perceived as saying that high-order schemes are necessarily oscillatory. Only much later was it realized that Godunov's result applies only to linear schemes and that it is possible to design *nonlinear* high order accurate schemes that are monotonicity preserving (see [1], [22], [6], [23], [7], [2], and [30]). Schemes of this type are the "modern shock-capturing schemes" referred to in the title of this paper.

In the rest of this paper we concentrate on the design and analysis of such highly nonlinear schemes. Even in the constant coefficient case these schemes are nonlinear to the extent that does not justify the use of local linear stability. Therefore, we shall start our journey into the nonlinear world by introducing the notion of total variation stability, which is more suitable to handle this type of schemes.

3.   Total Variation Stability and TVD Schemes

Glimm [4] has considered the numerical solution by a random choice method of an IVP for a system of conservation laws with initial data of small total variation, and proved existence of weak solutions by showing convergence of subsequences. Following ideas used in Glimm's convergence proof, we can formulate the following theorem for convergence to weak solutions.

Theorem 3.1:   Let v_h be a numerical solution of a conservative scheme (1.5).

(i)   If

(3.1) $$TV(v_h(\cdot,t)) \leqslant C \cdot TV(u_0)$$

where TV( ) denotes the total variation in x and C is a constant independent of h for $0 \leqslant t \leqslant T$, then any refinement sequence $h \longrightarrow 0$ with $\tau = 0(h)$ has a convergent subsequence $h_j \longrightarrow 0$ that converges in L_1^{loc} to a weak solution of (1.1).

(ii)   If v_h is consistent with an entropy inequality which implies uniqueness of the IVP (1.1), then the scheme is convergent (i.e., all subsequences have the same limit, which is the unique entropy solution of the IVP (1.1)).

We remark that unlike convergence theorems of classical numerical analysis, in which one shows that the distance between the solution and its numerical approximation vanishes as $h \longrightarrow 0$, the convergence argument in the above theorem relies on a combination of compactness and uniqueness; its relation to an existence proof is quite obvious (see [8] for more details).

We next demonstrate the use of Theorem 3.1 to prove convergence of schemes in conservation form (1.5) which are monotone, i.e., are of the form

$$(3.2) \qquad v_j^{n+1} = H(v_{j-k}^n,...,v_{j+k}^n) = (E_h \cdot v^n)_j,$$

where H is a monotone nondecreasing function of each of its arguments in the interval $[a,b]$, $a = \min v_j^0$, $b = \max v_j^0$. We note that the schemes of Godunov (1.7), Lax-Friedrichs (1.6) and the first order scheme (2.6) with $\theta \equiv 1$, are all monotone.

We start by observing that the operator E_h in (3.2) is order preserving

$$(3.3a) \qquad u \geqslant v \Rightarrow E_h \cdot u \geqslant E_h \cdot v.$$

Since E_h is also conservative,

$$(3.3b) \qquad \sum_j (E_h \cdot v)_j = \sum_j v_j$$

it follows then from a Lemma of Crandall and Tartar (see [3]) that E_h is ℓ_1-contractive, i.e., for all u and v in ℓ_1

$$(3.3c) \qquad \|E_h \cdot u - E_h \cdot v\|_{\ell_1} \leqslant \|u - v\|_{\ell_1}$$

(here $\|u\|_{\ell_1} = h \sum_j |u_j|$). Taking u to be a translate of v, i.e.,

$$u_j = v_{j+1} \qquad \text{for all } j$$

we get from (3.3c) that

$$(3.4a) \qquad TV(E_h \cdot v) \leqslant TV(v)$$

where

$$(3.4b) \qquad TV(w) \equiv \sum_j |w_{j+1} - w_j|.$$

It follows then that the numerical solution satisfies (3.1) with $C = 1$; thus we have established the convergence of subsequences. To show that all limit solutions are the same, we can use an argument of Barbara Keyfitz in the appendix to [11], which shows that (3.3c)

158

implies that the scheme is consistent with Oleinik's entropy condition. This shows that monotone schemes satisfy the requirements of Theorem 3.1 and thus are convergent.

Unfortunately, monotone schemes are necessarily only first order accurate (see [11]). However, once we give up the requirement (3.3a) that E_h be an order preserving operator and consider the larger class of schemes that satisfy only (3.4), i.e., schemes that are total-variation-diminishing (TVD), it becomes possible to obtain second order accuracy. Observe that TVD schemes are necessarily monotonicity preserving (see [7]).

The following theorem provides an almost complete characterization of TVD schemes (see [7], [8], and [18]).

Theorem 3.2: Let E_h be a numerical solution operator of the form

$$(3.5a) \qquad v_j^{n+1} = v_j^n + \sum_{\ell=-k}^{k-1} C_\ell(j)\Delta_{j-\ell-1/2}v^n \equiv (E_h \cdot v^n)_j.$$

where

$$(3.5b) \qquad \Delta_{i+1/2}v^n = v_{i+1}^n - v_i^n,$$

and $C_\ell(j)$ denotes some functional of v^n evaluated at j. Then E_h is TVD if (and only if)[1] the following relations hold:

$$(3.6a) \qquad C_{-1}(j-1) \geqslant C_{-2}(j-2) \ldots \geqslant C_{-k}(j-k) \geqslant 0$$

$$(3.6b) \qquad -C_0(j) \geqslant -C_1(j+1) \geqslant \ldots \geqslant -C_{k-1}(j+k-1) \geqslant 0$$

$$(3.6c) \qquad -C_0(j) + C_{-1}(j-1) \leqslant 1.$$

We turn now to consider the important case of $k = 1$ in (3.5), i.e.,

---

[1]Theorem 3.2 is not a complete characterization of TVD schemes, since the representation of a given nonlinear scheme in the form(3.5) is not unique.

(3.7) $$v_j^{n+1} = v_j^n + C_{-1}(j)\Delta_{j+1/2}v^n + C_0(j)\Delta_{j-1/2}v^n;$$

we refer to (3.7) as an essentially 3-point scheme, because the coefficients $C_0(j)$ and $C_{-1}(j)$ may depend on more than just $\{v_{j-1}^n, v_j^n, v_{j+1}^n\}$. To see the relation between the form (3.7) and the conservation form (1.5) let us consider the scheme

(3.8a) $$v_j^{n+1} = v_j^n - \lambda(\bar{f}_{j+1/2} - \bar{f}_{j-1/2})$$

with

(3.8b) $$\bar{f}_{i+1/2} = \frac{1}{2}(f_i + f_{i+1} - q_{i+1/2}\Delta_{i+1/2}v^n).$$

It is easy to see that (3.8) can be rewritten in the form (3.7) with

(3.9a) $$C_{-1}(j) = \frac{\lambda}{2}(\bar{a}_{j+1/2} + q_{j+1/2})$$

(3.9b) $$C_0(j) = \frac{\lambda}{2}(\bar{a}_{j-1/2} - q_{j-1/2});$$

here $\bar{a}_{i+1/2} = \bar{a}(v_i^n, v_{i+1}^n)$, which is defined by (1.8b).

Applying Theorem 3.2 to the scheme (3.8), we get that it is TVD if

(3.10a) $$\lambda|\bar{a}_{j+1/2}| \leq \lambda q_{j+1/2} \leq 1.$$

We turn now to outline the modified flux approach for the construction of second order accurate TVD schemes (see [7]). To simplify our presentation we choose in (3.10a)

(3.10b) $$q_{j+1/2} = |\bar{a}_{j+1/2}|;$$

this makes (3.8) identical to the Cole-Murman scheme (1.8). We observe that the TVD property of this scheme does not depend on the particular value of $f(u)$, but only on the CFL-like condition

160

(3.10c) $$\lambda |\bar{a}_{j+1/2}| \leqslant 1;$$

note that this condition involves only the *grid values* f_j. Consequently, if we apply this scheme to a modified flux $f_j^{mod} = f_j + g_j$, i.e.,

(3.11a) $$v_j^{n+1} = v_j^n - \lambda(\bar{f}_{j+1/2} - \bar{f}_{j-1/2})$$

(3.11b) $$\bar{f}_{j+1/2} = \frac{1}{2} [f_j + g_j + f_{j+1} + g_{j+1} - |\bar{a}_{j+1/2} + \bar{\gamma}_{j+1/2}| \Delta_{j+1/2} v^n]$$

where

(3.11c) $$\bar{\gamma}_{j+1/2} = (g_{j+1} - g_j)/\Delta_{j+1/2} v^n,$$

we can conclude that this scheme is TVD provided that

(3.12) $$\lambda |\bar{a}_{j+1/2} + \bar{\gamma}_{j+1/2}| \leqslant 1.$$

It is easy to verify by truncation error analysis that if

(3.13a) $$g_j = h\sigma(a)u_x + 0(h^2)$$

where

(3.13b) $$\sigma(x) = \frac{1}{2} |x| (1 - \lambda|x|),$$

then

(3.13c) $$\bar{f}_{j+1/2} = \bar{f}_{j+1/2}^{LW} + 0(h^2),$$

where $\bar{f}^{LW}$ is the numerical flux (1.9) of the second-order accurate Lax-Wendroff scheme.

In [7] we have taken g_j to be

(3.14a) $$g_j = m(\sigma(\bar{a}_{j-1/2})\Delta_{j-1/2}v^n, \ \sigma(\bar{a}_{j+1/2})\Delta_{j+1/2}v^n)$$

161

where

$$(3.14b) \quad m(x,y) = \begin{cases} s \cdot \min(|x|,|y|) & \text{if } sgn(x) = sgn(y) = s \\ 0 & \text{otherwise} \end{cases}$$

Clearly g_j (3.14a) satisfies (3.13a) and consequently the resulting scheme is second-order accurate, except at local extreme where the $O(h^2)$ term in (3.13a) and (3.13c) fails to be Lipschitz continuous.

Next we show that due to this particular definition of g_j, the modified flux scheme (3.11) which is second-order accurate, is also TVD under the original CFL restriction (3.10c); this follows immediately from the following lemma.

Lemma 3.3:

$$(3.15a)(i) \qquad |\bar{\gamma}_{j+1/2}| \leq 2|\sigma(\bar{a}_{j+1/2})|$$

$$(3.15b)(ii) \qquad \lambda|\bar{a}_{j+1/2}| \leq 1 \Rightarrow \lambda|\bar{a}_{j+1/2} + \bar{\gamma}_{j+1/2}| \leq 1.$$

Proof: We note that $m(x,y)$ (3.14b) satisfies $|m(x,y)| \leq \min(|x|,|y|)$. Consequently

$$|g_{j+1}-g_j| \leq |g_j| + |g_{j+1}| \leq \min(|\sigma_{j-1/2}\Delta_{j-1/2}v^n|, |\sigma_{j+1/2}\Delta_{j+1/2}v^n|)$$

$$+ \min(|\sigma_{j+1/2}\Delta_{j+1/2}v^n|, |\sigma_{j+3/2}\Delta_{j+3/2}v^n|)$$

$$\leq 2|\sigma(\bar{a}_{j+1/2})||\Delta_{j+1/2}v^n|,$$

which proves (3.15a).

It follows therefore from (3.13b) and (3.15a) that

$$\lambda|\bar{a} + \bar{\gamma}| \leq \lambda|\bar{a}| + \lambda|\bar{\gamma}| \leq \lambda|\bar{a}| + 2\lambda|\sigma(\bar{a})|$$

$$= \lambda|\bar{a}| + \lambda|\bar{a}|(1 - \lambda|\bar{a}|) \leq \lambda|\bar{a}| + 1 - \lambda|\bar{a}| = 1,$$

which proves this lemma.

162

We remark that the modified flux scheme (3.11), as the Cole-Murman scheme it is derived from, admits a stationary "expansion shock" as a steady solution. Replacing $q_{j+1/2} = |\bar{a}_{j+1/2}|$ in (3.10b) by

$$(3.16) \qquad q_{j+1/2} = \max(|\bar{a}_{j+1/2}|, \epsilon/\lambda), \qquad \epsilon > 0$$

results in a modified flux scheme which is entropy consistent (see [28]) and thus can be shown to be convergent by Theorem 3.1.

The choice (3.14) of g_j is by no means unique. It is easy to check that changing g_j to be

$$(3.14a)' \qquad g_j = \bar{m}(\sigma(\bar{a}_{j-1/2})\Delta_{j-1/2}v^n, \sigma(\bar{a}_{j+1/2})\Delta_{j+1/2}v^n)$$

with

$$(3.14b)' \qquad \bar{m}(x,y) = \begin{cases} x & \text{if } |x| < |y| \\ y & \text{if } |x| \geq |y| \end{cases}$$

or

$$(3.14a)'' \qquad g_j = m(\ (1-\lambda|\bar{a}_{j-1/2}|)\Delta_{j-1/2}v^n, \frac{1}{2}\sigma(a_j)(v^n_{j+1} - v^n_{j-1}),$$

$$(1 - \lambda|\bar{a}_{j+1/2}|)\Delta_{j+1/2}v^n)$$

with

$$(3.14b)'' \qquad m(x,y,z) = m(m(x,y),z),$$

does not alter the relations (3.13a) and (3.15a) which makes the modified flux scheme (3.11) a second-order accurate TVD scheme, under the original CFL restriction (3.10c).

The modified flux approach is not the only methodology to construct second-order accurate TVD schemes (there are many ways to skin a nonlinear cat). In the next section, we shall describe the MUSCL scheme of van Leer [23]; other techniques are described in [30], [27], and [31]. Unfortunately, all TVD schemes, independent of their derivation, are only first order accurate at local extrema of

163

the solution. Consequently, TVD schemes can be second-order accurate in the L_1 sense, but only first order accurate in the maximum norm (see [14] for more details).

4.   Godunov-Type Schemes

In this section we describe Godunov-type schemes which are an abstraction of Godunov's scheme (1.7) (see [5]) due to ideas in [23], [12], and [13].

We start with some notations: Let $\{I_j\}$ be a partition of the real line; let $A(I)$ denote the interval-averaging (or "cell-averaging") operator

$$(4.1) \qquad A(I) \cdot w = \frac{1}{|I|} \int_I w(y)dy;$$

let $\bar{w}_j = A(I_j) \cdot w$ and denote $\bar{w} = \{\bar{w}_j\}$. We denote the approximate reconstruction of $w(x)$ from its given cell-averages $\{\bar{w}_j\}$ by $R(x;\bar{w})$. To be precise, $R(x;\bar{w})$ is a piecewise-polynomial function of degree $(r-1)$, which satisfies

$$(4.2a)(i) \qquad R(x;\bar{w}) = w(x) + O(h^r) \qquad \text{wherever w is smooth}$$

$$(4.2b)(ii) \qquad A(I_j) \cdot R(\cdot;\bar{w}) = \bar{w}_j \text{ (conservation).}$$

Finally, we define Godunov-type schemes by

$$(4.3a) \qquad v_j^{n+1} = A(I_j^{n+1}) \cdot E(\tau) \cdot R(\cdot;v^n) \equiv (\bar{E}_h \cdot v^n)_j$$

$$(4.3b) \qquad v_j^0 = A(I_j^0)u_0;$$

here $\{I_j^n\}$ is the partition of the real line at time t_n, and $E(t)$ is the evolution operator of (1.1).

In the scalar case, both the cell-averaging operator $A(I_j)$ and the solution operator $E(\tau)$ are order-preserving, and consequently also total-variation diminishing (TVD); hence

$$(4.4) \qquad TV(\bar{E}_h \cdot \bar{w}) \leqslant TV(R(\cdot;\bar{w})).$$

This shows that the total variation of the numerical solution of Godunov-type schemes is dominated by that of the reconstruction step.

165

The original first-order accurate scheme of Godunov is (4.3) with the piecewise-constant reconstruction

(4.5) $R(x;\bar{w}) = \bar{w}_j,$ for $x \in I_j$.

Since the piecewise-constant reconstruction (4.5) is an order-preserving operation, it follows that $\bar{E}_h$ is likewise order preserving as a composition of 3 such operations; consequently the scheme is monotone.

The second-order accurate MUSCL scheme of van Leer [23] is (4.3) with the piecewise-linear reconstruction

(4.6a) $R(x;\bar{w}) = \bar{w}_j + (x - y_j) \cdot s_j$ for $x \in I_j$,

where s_j is defined by

(4.6b) $s_j = m(\Delta_{j-1/2}\bar{w}/\Delta_{j-1/2}y, \Delta_{j+1/2}\bar{w}/\Delta_{j+1/2}y);$

here y_j denotes the center of I_j. It is easy to verify that the particular form of the slope s_j in (4.6) implies that

(4.7a) $TV(R(\cdot;\bar{w})) = TV(\bar{w});$

hence it follows from (4.4) that the scheme is TVD, i.e.,

(4.7b) $TV(\bar{E}_h \cdot \bar{w}) \leqslant TV(\bar{w}).$

To simplify our presentation, we assume from now on that the partition $\{I_j^n\}$ is stationary and uniform, i.e.

(4.8) $I_j^n = (x_{j-1/2}, x_{j+1/2});$

this enables us to express the schemes (4.3) by standard grid notations.

The Godunov-type scheme (4.3) generates discrete values

166

$\{v_j^n\}$, which are r-th order accurate approximations to $\{\bar{u}_j^n\}$, the cell-averages of the exact solution. We note, however, that the operation of the scheme (4.3) also involves a globally defined pointwise approximation to u(x,t) of the same order of accuracy which we denote by $v_h(x,t)$. The latter is defined for all x in the time-strips $t_n \leqslant t < t_{n+1}$ by

(4.9) $\qquad v_h(\cdot,t_n+t) = E(t) \cdot R(\cdot;v^n) \qquad$ for $\qquad 0 \leqslant t < \tau$.

We remark that (4.3) is the abstract operator expression of a scheme in the standard conservation form

(4.10a) $\qquad v_j^{n+1} = v_j^n - \lambda(\bar{f}_{j+1/2} - \bar{f}_{j-1/2})$

with the numerical flux

(4.10b) $\qquad \bar{f}_{j+1/2} = \dfrac{1}{\tau} \int\limits_0^\tau f(v_h(x_{j+1/2},t_n+t))dt.$

For r = 1 (Godunov's scheme), the numerical flux (4.10b) can be expressed by (1.7). For $r \geqslant 2$, we make use of the fact that $v_h(x_{j+1/2},t_n+t)$ in (4.10b) is needed only "in the small", in order to derive simple but adequate approximations to the numerical flux (see [16] for more details).

We remark that (4.7a) is sufficient but not a necessary condition for the scheme $\bar{E}_h$ to be TVD (4.7b). Other choices of the slope s_j in (4.6), such as

(4.6b)' $\qquad h \cdot s_j = \bar{m}(\Delta_{j-1/2}\bar{w},\Delta_{j+1/2}\bar{w})$

or

(4.6b)'' $\qquad h \cdot s_j = m(2\Delta_{j-1/2}\bar{w},(\bar{w}_{j+1}-\bar{w}_{j-1})/2,2\Delta_{j+1/2}\bar{w})$

do not satisfy (4.7a); nevertheless the resulting scheme is TVD. This

167

is due to the helping hand of the cell-averaging operator, which is not taken into account in (4.4).

MUSCL-type schemes, as all other TVD schemes, are second-order accurate only in the L_1-sense. In order to achieve higher-order of accuracy, we have to weaken our control over the possible increase in total variation due to the reconstruction step. We do so by introducing the notion of essentially non-oscillatory (ENO) schemes in the next section.

168

5.   ENO Schemes

We turn now to describe the recently developed essentially non-oscillatory (ENO) schemes of [16], which can be made accurate to any finite order r. These are Godunov-type schemes (4.3) in which the reconstruction R(x;$\bar{w}$), in addition to relations (4.2), also satisfies

(5.1) $TV(R(\cdot;\bar{w})) \leqslant TV(\bar{w}) + O(h^{1+p})$, $p > 0$

for any piecewise-smooth function w(x).   Such a reconstruction is essentially non-oscillatory in the sense that it may not have a Gibbs-like phenomenon at jump-discontinuities of w(x), which involves the generation of O(1) spurious oscillations (that are proportional to the size of the jump); it can, however, have $small$ spurious oscillations which are produced in the smooth(er) part of w(x), and are usually of the size $O(h^r)$ of the reconstruction error (4.2a).

When we use an essentially non-oscillatory reconstruction in a Godunov-type scheme, it follows from (4.4) and (5.1) that the resulting scheme (4.3) is likewise essentially non-oscillatory (ENO) in the sense that for all piecewise-smooth functions w(x)

(5.2) $TV(\bar{E}_h \cdot \bar{w}) \leqslant TV(\bar{w}) + O(h^{1+p})$, $p > 0$;

i.e., it is "almost TVD". Property (5.2) makes it reasonable to believe that at time t = T, after applying the scheme $N = T/\tau = O(h^{-1})$ times, we can expect

(5.3) $TV(v_h(\cdot,T)) \leqslant C \cdot TV(u_0) + O(h^p)$.

We recall that by Theorem 3.1, this implies that the scheme is convergent (at least in the sense of having convergent subsequences). This hope is supported by a very large number of numerical experiments. In order to conclude from (5.2) that for all n $\geqslant$ 0,

(5.3) $TV(v^{n+1}) \leqslant TV(v^n) + O(h^{1+p})$, $p > 0$

169

we still have to show that, starting from a piecewise-smooth $u_0(x)$ in (4.3b), v^n remains sufficiently close in its regularity to a piecewise-smooth function, so that (5.2) applies to the following time-steps as well. Unfortunately, we have not been able as yet to analyze the regularity of v^n.

Next we describe one of the techniques to obtain an ENO reconstruction. Given cell-averages $\{\bar{w}_j\}$ of a piecewise smooth function $w(x)$, we observe that

$$(5.4a) \qquad h\bar{w}_j = \int_{x_{j-1/2}}^{x_{j+1/2}} w(y)dy = W(x_{j+1/2}) - W(x_{j-1/2})$$

where

$$(5.4b) \qquad W(x) = \int_{x_0}^{x} w(y)dy$$

is the primitive function of $w(x)$. Hence we can easily compute the *point values* $\{W(x_{i+1/2})\}$ by summation

$$(5.4c) \qquad W(x_{i+1/2}) = h \sum_{j=i_0}^{i} \bar{w}_j.$$

Let $H_m(x;u)$ be an interpolation of u at the points $\{y_j\}$, which is accurate to order m, i.e.

$$(5.5a) \qquad H_m(y_j;u) = u(y_j),$$

$$(5.5b) \qquad \frac{d^\ell}{dx^\ell} H_m(x;u) = \frac{d^\ell}{dx^\ell} u(x) + O(h^{m+1-\ell}), \qquad 0 \leqslant \ell \leqslant m.$$

We obtain our "reconstruction via primitive function" technique by defining

(5.6)
$$R(x;\bar{w}) = \frac{d}{dx} H_r(x;W).$$

Relation (4.2a) follows immediately from (5.5b) with $\ell = 1$ and the definition (5.4), i.e.,

$$R(x;\bar{w}) = \frac{d}{dx} H_r(x;W) = \frac{d}{dx} W(x) + O(h^r)$$

$$= w(x) + O(h^r)$$

Relation (4.2b) is a direct consequence of (5.5a) and (5.4), i.e.,

$$A(I_j)R(\cdot;\bar{w}) = \frac{1}{h} \int_{x_{j-1/2}}^{x_{j+1/2}} \frac{d}{dx} H_r(x,W)dx$$

$$= \frac{1}{h} [H_r(x_{j+1/2};W) - H_r(x_{j-1/2};W)]$$

$$= \frac{1}{h} [W(x_{j+1/2}) - W(x_{j-1/2})] = \bar{w}_j.$$

To obtain an ENO reconstruction, we take H_r in (5.6) to be the new ENO interpolation technique of the author [9]. In this case, $H_m(x;u)$ is a piecewise–polynomial function of x of degree m, which is defined (omitting the u dependence) by

(5.7a)
$$H_m(x;u) = q_{j+1/2}(x) \qquad \text{for} \qquad y_j \leqslant y \leqslant y_{j+1}$$

where $q_{j+1/2}$ is the unique polynomial of degree m that interpolates u at the m+1 points

(5.7b)
$$S_m(i) \equiv \{y_i,...,y_{i+m}\}$$

for a particular choice of $i = i(j)$ (to be described in the following). To satisfy (5.5a), we need

$$q_{j+1/2}(y_j) = u(y_j), \ q_{j+1/2}(y_{j+1}) = u(y_{j+1});$$

therefore, we limit our choice of i(j) to

(5.7c) $$j - m+1 \leq i(j) \leq j.$$

The ENO interpolation technique is nonlinear: At each interval $[y_j, y_{j+1}]$, we consider the m possible choices of stencils (5.7b) subject to the restriction (5.7c), and assign to this interval the stencil in which u is "smoothest" in some sense; this is done by specifying i(j) in (5.7b).

The information about the smoothness of u can be extracted from a table of divided differences. The k-th divided difference of u

(5.8a) $$u[y_i, y_{i+1}, \ldots, y_{i+k}] \equiv u[S_k(i)]$$

is defined inductively by

(5.8b) $$u[S_0(i)] = u(y_i)$$

and

(5.8c) $$u[S_k(i)] = (u[S_{k-1}(i+1)] - u[S_{k-1}(i)])/(y_{i+k} - y_i).$$

If u(x) is m times differentiable in $[y_i, y_{i+m}]$ then

(5.9a) $$u[S_m(i)] = \frac{1}{m!} u^{(m)}(\xi), \text{ for some } y_i \leq \xi \leq y_{i+m}$$

If $u^{(p)}(x)$ has a jump discontinuity in $[y_i, y_{i+m}]$ then

(5.9b) $$u[S_m(i)] = O(h^{-m+p}[u^{(p)}]), \ 0 \leq p \leq m-1$$

($[u^{(p)}]$ in the RHS of (5.9b) denotes the jump in the p-th derivative).

Relations (5.9) show that $|u[S_m(i)]|$ is a measure of the smoothness of u in $S_m(i)$, and therefore can serve as a tool to compare the relative smoothness of u in various stencils. The simplest algorithm to assign $S_m(i(j))$ to the interval $[y_j, y_{j+1}]$ is the following:

172

Algorithm I: Choose i(j) so that

(5.10) $$|u[S_m(i(j))]| = \min_{j-m+1 \leqslant i \leqslant j} \{|u[S_m(i)]|\}.$$

Clearly (5.10) selects the "smoothest" stencil, provided that h is sufficiently small (but not smaller than the round-off error of the machine would permit!).

In order to make a sensible selection of stencil also in the "pre-asymptotic" case, we prefer to use the following hierarchial algorithm:

Algorithm II: Let $i_k(j)$ be such that $S_k(i_k(j))$ is our choice of a (k+1)-point stencil for $[y_j, y_{j+1}]$. Obviously we have to set

(5.11a) $$i_1(j) = j$$

To choose $i_{k+1}(j)$, we consider as candidates the two stencils

(5.11b) $$S^L_{k+1} = S_{k+1}(i_k(j) - 1),$$

(5.11c) $$S^R_{k+1} = S_{k+1}(i_k(j)),$$

which are obtained by adding a point to the left of (or to the right of) $S_k(i_k(j))$, respectively.   We select the one in which u is relatively smoother, i.e.,

(5.11d) $$i_{k+1}(j) = \begin{cases} i_k(j) - 1 & \text{if } |u[S^L_{k+1}]| < |u[S^R_{k+1}]| \\ i_k(j) & \text{otherwise.} \end{cases}$$

Finally we set $i(j) = i_m(j)$.

Using Newton's form of interpolation, we see that the polynomials $\{q_k(x)\}$, $1 \leqslant k \leqslant m$, corresponding to the stencils $S^k = S_k(i_k(j))$ selected by Algorithm II, satisfy the relation

(5.11e) $$q_{k+1}(x) = q_k(x) + u[S^{k+1}] \prod_{y \in S^k} (x-y).$$

173

This shows that the choice made in (5.11d) selects q_{k+1} to be the one that deviates the least from q_k. It is this property that makes Algorithm II meaningful also for h in the pre-asymptotic range.

In Figure 3, we apply the piecewise polynomial interpolation (5.7) to a piecewise-smooth function u which has in [-1,1] 3 jump discontinuities in the function itself and another one in the first derivative. This function is shown in Figure 3 by a continuous line on which there are 30 circles that denote the values used for the interpolation. This function was continued periodically outside [-1,1] (not shown in the picture).

In Figure 3a, we show the 6-th order polynomial (5.7) (i.e., m = 6) with the *predetermined* stencil i(j) = j; i.e., the 7-points stencil $\{y_j, y_{j+1}, ..., y_{j+6}\} = S_6(j)$ is used to define $q_{j+1/2}$ in $[y_j, y_{j+1}]$. Figure 3a shows a highly oscillatory behavior of the interpolation polynomial.

In Figure 3b, we show the same 6-th order polynomial (5.7) except that now we use the *adaptive* stencil which is selected by Algorithm II (5.11).

To understand why this interpolation works as well as it does, we consider the following two possibilities:

(i) $[y_j, y_{j+1}]$ is in the smooth part of u: For h sufficiently small, both Algorithms I and II choose a stencil $S_m(i(j))$ which is also in the smooth part of the function. In this case, (5.5b) in $[y_j, y_{j+1}]$ is the standard result for m-th order interpolation of a smooth function. We observe that $q_{j+1/2}$ need not be a monotone approximation to u in $[y_j, y_{j+1}]$; nevertheless, its total variation there cannot be more than $O(h^{m+1})$ larger than that of u.

(ii) $[y_j, y_{j+1}]$ contains a discontinuity: For h sufficiently small, the function u near $[y_j, y_{j+1}]$ can be thought of as a step-function. In the case of a step-function, the particular choice of i(j) is of no importance since all the stencils $S_m(i)$ with $j-m+1 \leqslant i \leqslant j$ lead to a $q_{j+1/2}(x)$ which is montone in (y_j, y_{j+1}). This follows from the simple observation that in the case of a

174

step-function, we have for all $1 \leqslant \ell \leqslant m$, except $\ell = j-i$

(5.12a) $$u(y_{i+\ell}) = u(y_{i+\ell+1}),$$

and, consequently, also

(5.12b) $$q_{j+1/2}(y_{i+\ell}) = q_{j+1/2}(y_{i+\ell+1}).$$

Using Rolle's theorem, we count in (5.12b) $(m-1)$ roots of $dq_{j+1/2}/dx$ outside (y_j, y_{j+1}). Since $dq_{j+1/2}/dx$ is a polynomial of degree $(m-1)$, it follows that these are all its roots. Hence, $dq_{j+1/2}/dx$ does not vanish in (y_j, y_{j+1}), which shows that it is monotone there (see [17] and [15] for more details).

We conclude this section by showing in Figures 4 and 5 the solution to the shock-tube problem (2.7) by the ENO scheme with $r = 2$ (Figure 4) and $r = 4$ (Figure 5). Comparing Figures 4-5 to Figures 1-2, we observe a considerable improvement in performance (see [14] for more details).

6. Nonlinearity, Upwind Differencing and Linear Stability

In this section, we consider the constant coefficient case (2.1). In this case, the Godunov-type scheme (4.3) can be expressed as

$$(6.1a) \qquad v_j^{n+1} = \bar{R}(x_j - a\tau; v^n),$$

where $\bar{R}(x; \cdot)$ denotes the sliding average of R, i.e.,

$$(6.1b) \qquad \bar{R}(x; \cdot) = \frac{1}{h} \int_{-h/2}^{h/2} R(x+\xi; \cdot)d\xi.$$

We note that since R is a piecewise polynomial of degree $(r-1)$, $\bar{R}$ is a piecewise-polynomial of degree r. Moreover, the conservation property (4.2b) shows that $\bar{R}(x; v^n)$ is an interpolation of $\{v_j^n\}$. It is interesting to note that using R which is obtained via the primitive function (5.6), we get from (6.1) the particularly simple form

$$(6.2a) \qquad v_j^{n+1} = \frac{1}{h} [H_r(x_{j+1/2} - a\tau; V^n) - H_r(x_{j-1/2} - a\tau; V^n)],$$

where $\{V_{j+1/2}^n\}$ is defined at $\{x_{j+1/2}\}$ by (5.4), i.e.,

$$(6.2b) \qquad V_{j+1/2}^n = h \sum_{i=i_0}^{j} v_i^n.$$

Relation (6.2) directly relates Godunov-type schemes to interpolation. Clearly, if the interpolation H_r is based on a fixed stencil, then the resulting scheme is linear; the non-linearity of the ENO schemes stems from its adaptive selection of stencil.

When $r = 1$, $\bar{R}$ in (6.1b) is necessarily the piecewise-linear interpolation of $\{v_j^n\}$; consequently the "upwind shift" $(-a\tau)$ forces the scheme to be the first-order upwind scheme. We recall however that the stencil in the ENO scheme is chosen from considerations of smoothness which have nothing to do with the PDE; the "upwind shift" $(-a\tau)$ is only by one cell; consequently the resulting ENO scheme (6.2) for $r \geq 2$ need not be, and in general is not, "upwind".

We turn now to study the second-order accurate ENO scheme

176

$(r = 2$ in $(6.1) - (6.2))$. It is easy to see that this scheme is identical to the MUSCL-type scheme (4.6) with s_j defined by $(4.6b)'$. It is somewhat more surprising to find that the MUSCL-type scheme (in the constant coefficient case), is identical to the second-order accurate modified-flux scheme (3.11) with the correspondence

$$(6.3) \qquad g_j = h\sigma(a)s_j.$$

Consequently, all these second-order accurate TVD schemes can be written as (6.1) with a piecewise-parabolic $\bar{R}(x;v^n)$. For $a < 0$, we get

$$v_j^{n+1} = \bar{R}(x_j - a\tau; v^n)$$
$(6.4a)$
$$= v_j^n - \lambda a \Delta_{j+1/2} v^n + 1/2 a(1+\lambda a) \cdot (s_{j+1} - s_j),$$

which is obtained by taking the sliding-average of $(4.6a)$.

We observe that when in $(6.4b)$

$$(6.4b) \qquad h \cdot s_j = \Delta_{j+1/2} v^n, \quad h \cdot s_{j+1} = \Delta_{j+3/2} v^n$$

then $(6.4a)$ is the second-order accurate upwind-differencing scheme. However, when in $(6.4b)$

$$(6.4c) \qquad h \cdot s_j = \Delta_{j-1/2} v^n, \quad h \cdot s_{j+1} = \Delta_{j+1/2} v^n$$

then $(6.4a)$ is the central-differencing Lax-Wendroff scheme.

Based on this observation, we see that the MUSCL-type scheme with s_j defined by $(4.6b)$ or $(4.6b)'$ satisfies $(6.4b)$ when, as a function of i,

$$(6.4b)' \qquad \{ |\Delta_{i+1/2} v^n| \} \text{ is decreasing,}$$

and it satisfies $(6.4c)$ when

177

(6.4c)' $\{|\Delta_{i+1/2}v^n|\}$ is increasing.

This shows that the "popular" reference to the MUSCL scheme and the modified flux scheme as "upwind differencing" schemes is not justified.

We remark that the scheme (6.4a) is second-order accurate only if

(6.5a) $$s_{j+1} - s_j = hu_{xx} + O(h^2).$$

This shows that in addition to

(6.5b) $$s_j = u_x(x_j) + O(h)$$

we need also the Lipschitz-continuity of the $O(h)$ term in (6.5b). As we have mentioned earlier, the MUSCL scheme, as well as the modified flux scheme and other TVD schemes, fail to have this extra smoothness at local extrema, which are the transition points between (6.4b) and (6.4c); consequently, their accuracy drops to first order at points of local extremum.

The analysis of these second-order accurate nonlinear schemes shows that the "nature" of the scheme depends on differences of its numerical solution; therefore, local linearization is not justified. Since the two schemes in (6.4) are linearly stable, such incorrect linearization would nevertheless result in a correct statement of stability. This is not the case for $r > 2$, where, as r increases, more and more of the various choices of stencil can be identified as if belonging to a linearly unstable scheme. Since Fourier analysis is valid only if the same stencil is used everywhere, this identification is not necessarily relevant and may actually be quite misleading.

A situation of this type is encountered when we consider the initial-boundary-value problem (IBVP) in $-1 \leqslant x \leqslant 1$

(6.6) $$u_t + u_x = 0, \; u(x,0) = u_0(x), \; u(-1,t) = g(t);$$

$x = 1$ is a "outflow boundary" and no condition needs to be specified there. We divide $[-1,1]$ into $(J+1)$ interval $\{I_j\}_{j=0}^{J}$, where $I_j = (x_{j-1/2}, x_{j+1/2})$ and

(6.7a) $$x_{-1/2} = -1, \; x_{J+1/2} = 1.$$

Given cell averages $\{\bar{w}_j\}$ for $j = 0,...,J$ we define $W(x_{-1/2}) = 0$ and compute $W(x_{j+1/2})$, $j = 1,...,J$ by (5.4c) with $i_0 = 0$; thus $W(x_{j+1/2})$ is given also at $x = \pm 1$. Next we evaluate $H_r(x;w)$ by Algorithm II, which is modified so that the choice of stencil in (5.11) is restricted to available data. Thus, $H_r(x;w)$ is defined for $-1 \leqslant x \leqslant 1$, and as before we define

(6.6b) $$R(x;\bar{w}) = \frac{d}{dx} H_r(x;w), \quad -1 \leqslant x \leqslant 1.$$

Using this definition of $R(x;\bar{w})$ in $[-1,1]$, we modify the Godunov-type scheme (4.3) by

(6.7a) $$v_j^{n+1} = A(I_j) \cdot \tilde{E}(\tau) \cdot R(\cdot;v^n), \quad 0 \leqslant j \leqslant J.$$

(6.7b) $$v_j^0 = A(I_j) \cdot u_0.$$

Here $\tilde{v}(t) = \tilde{E}(t) \cdot R(\cdot;v^n)$ is the solution in the small (i.e., for $0 \leqslant t \leqslant \tau$) of the IBVP

(6.7c) $$\tilde{v}_t + f(\tilde{v})_x = 0, \; \tilde{v}(x,0) = R(x;v^n), \; \tilde{v}(-1,t) = g(t_n+t).$$

This implementation of Godunov-type schemes to IBVP's is very convenient: There are no "artificial numerical boundaries", and the prescribed boundary conditions are handled on the level of the PDE (6.7c). We observe, however, that near $x = -1$ the scheme is "differenced against the wind", which is linearly unstable if done everywhere. Therefore, our experience with linear schemes may inhibit us from using this approach. Overcoming this inhibition, we have performed a large number of numerical experiments with the so modified ENO schemes (two of which are presented in the following) and we are happy to report that these schemes have been found to be

179

stable in all our experiments.

In Table 1, we present a mesh-refinement chart for the IBVP (6.6) with

(6.8) $u(x,0) = \sin \pi x, \; u(-1,t) = -\sin \pi(1+t)$.

The ENO schemes were used with a CFL number of 0.8, and the results are shown at $t = 2$. Table 1 indicates that the ENO schemes with $1 \leqslant r \leqslant 6$ are convergent in this case; the accumulation of error seems to be linear. Comparing Table 1 to the periodic case (see [9]), we observe that the results for the IBVP are slightly better in the asymptotic range, which is to be expected.

Next we consider the IBVP (6.6) with

(6.9a) $u(x,0) = e^{-x}, \; u(-1,t) = e^{1+t}$,

the solution to which is

(6.9b) $u(x,t) = e^{-x+t}$.

We observe that: (i) $|u^{(k)}(x,t)|$ is a monotone decreasing function of x for all k and t. Consequently, if we apply Algorithm II to $\bar{u}(\cdot,t)$ we get $i(j) = j$ in (5.11). (ii) The scheme (6.2) with the *fixed* choice $i(j) = j$ is linear and strongly "biased against the wind"; consequently, it is linearly unstable.

In Table 2, we present a mesh-refinement chart for the solution at $t = 1$ of the IBVP (6.9) by the 4-th order ENO scheme ($r = 4$ in (6.6) – (6.7)) with CFL = 0.4. In spite of the previous observations, we find that the scheme seems to be convergent. This "paradox" is resolved once we examine the data in Figures 6 and 7 for $J = 80$ and 160, respectively. In (a), (b), (c) and (d), we show the absolute value of the k-th divided difference $|v[S_k]|$ for $k = 0,1,2,3$, respectively. We see that the numerical solution and its first divided difference are monotone. However, the second and third divided differences are oscillatory. This allows the scheme to select $i(j) \neq j$ in (5.11). The actual choice of $i(j)$ at $t = 1$ is shown in Figure 6e and Figure 7e; the straight line in these figures is i(j) = j. Comparing Figure 6d to Figure 7d, we see that the oscillations in $v[S_3]$, the third divided difference of v, are uniformly bounded under refinement. Analysis of the numerical data suggests that

(i) $v[S_3] \xrightarrow{\ w\ } u^{(3)}(x,t)$ (in an average sense) as $h \longrightarrow 0$;

(ii) $v[S_k] = u^{(k)}(x,t) + 0(h^{3-k})$ for $k = 0,1,2$.

Finally, we consider the application of the 4-th order ENO scheme to the periodic IVP

(6.10a) $u_t + u_x = 0,\ u(x,0) = ???,\ v_j^0 = (-1)^j.$

We observe that the mesh oscillation data in (6.10a),

(6.10b) $v_j^0 = (-1)^j = e^{ij\Pi}$

is the highest frequency in (2.3b), which determines the linear stability of the constant coefficient scheme (2.2). We note, however, that as h decreases, the total variation of v^0 becomes unbounded. Consequently, v^0 does not represent a BV function and, therefore, need not be considered when testing for total-variation-stability in (3.1). In the following, we describe numerical experiments where we apply the 4-th order ENO scheme to (6.10) anyhow. The selection of stencil (5.11) is designed to make a sensible choice only when applied to piecewise-smooth data. In the mesh-oscillation case $|v[S_k(i)]|$ is constant as a function of i for all k; consequently, (5.11) results in the arbitrary choice of the uniform stencil i(j) = j-3 (see Figure 8b). As in the previous case, the ENO scheme becomes a constant coefficient scheme (2.2) for which linear stability analysis applies. In Figure 8c, we show the amplification factor of the mesh-oscillation mode

(6.10c) $g(\nu) = \sum_{\ell=-k}^{k} (-1)^\ell C_\ell(\nu)$

181

as a function of the CFL number $\nu = \lambda a$. The amplification factor (6.10c) for the ENO schemes is determined by two competing factors: (i) Increase of oscillations due to the reconstruction, which is based on the highly-oscillatory interpolation of the mesh oscillation (6.10b); (ii) Decrease of oscillations due to the operation of cell-averaging on the translated data. Figure 8c shows that for $\nu < 0.26$, the latter wins and the scheme is linearly stable; for larger values of ν the scheme is linearly unstable. In Figures 8d and 8e, we show the numerical solution of the 4-th order ENO scheme with $\nu = 0.6$ after a single time-step (n=1) and twenty time-steps (n=20), respectively. Clearly, the numerical solution blows up like $(1.67)^n$.

It is amusing to realize that this "linear instability" is itself "nonlinearly unstable" in the sense that any nonuniform perturbation of the mesh-oscillation data turns the ENO scheme into a stable nonlinear scheme. To demonstrate this point, we perturb the mesh-oscillation data by a random noise of the size 10^{-6} of the round-off error (see Figures 9a and 9b), and repeat the previous calculation. In Figures 9d-9k, we present subsequent "snapshots" of the numerical solution, which show that the numerical solution decays in both the amplitude and the number of oscillations; observe that the rate of decay is faster for the highly oscillatory components of the solution and slower for the smoother ones.

This property enables the scheme to combine "robustness" with accuracy. We demonstrate this feature of the ENO schemes in Figure 10 where we apply the 4-th order scheme with $\nu = 0.4$ to initial data of sin πx perturbed by random noise of the size 10^{-1}; the squares denote the numerical solution; the continuous line shows sin πx.

Acknowledgement

I would like to thank Sukumar Chakravarthy, Bjorn Engquist and Stan Osher for various contributions to this research, and for making my stay at UCLA the pleasant and fruitful period that it was.

This research was supported by NSF Grant No. DMS85-03294,

ARO Grant No. DAAG29–86–K0190, and NASA Consortium Agreement no. MCA2–IR390–403. The author was supported by NASA Contracts NAS1–17070 and NAS1–18107 while in residence at ICASE.

References

[1]   Boris, J.P. and D.L. Book, "Flux corrected transport. I. SHASTA, a fluid transport algorithm that works," J. Computational Phys., v. 11, 1973, pp. 38-69.

[2]   Colella, P. and P.R. Woodward, "The piecewise-parabolic method (PPM) for gas-dynamical simulations," J. Comp. Phys., v. 54 (1984), 174-201.

[3]   Crandall, M.G. and A. Majda, "Monotone difference approximations for scalar conservation laws," Math. Comp., 34(1980), pp. 1-21.

[4]   Glimm, J., "Solutions in the large for nonlinear hyperbolic systems of equation," Comm. Pure Appl. Math., 18(1965), pp. 697-715.

[5]   Godunov, S.K., "A difference scheme for numerical computation of discontinuous solutions of equations of fluid dynamics," Math. Sbornik, 47(1959), pp. 271-306. (In Russian.)

[6]   Harten, A., "The artificial compression method for computation of shocks and contact-discontinuities: III. Self-adjusting hybrid schemes," Math. Comp., v. 32(1978), pp. 363-389.

[7]   Harten, A., "High resolution schemes for hyperbolic conservation laws," J. Comp. Phys., v. 49(1983), pp. 357-393.

[8]   Harten, A., "On a class of High Resolution Total-Variation-Stable Finite-Difference Schemes," SINUM, v. 21(1984), pp. 1-23.

184

[9]   Harten, A., "On high-order accurate interpolation for
        non-oscillatory shock capturing schemes," MRC Technical
        Summary Report #2829, University of Wisconsin, (1985).

[10]   Harten, A. and J.M. Hyman, "A self-adjusting grid for the
        comptuation of weak solutions of hyperbolic conservation
        laws," J. Comp. Phys., v. 50(1983), pp. 235-269.

[11]   Harten, A., J.M. Hyman and P.D. Lax, "On finite-difference
        approximations and entropy conditions for shocks," Comm. Pure
        Appl. Math., 29(1976), pp. 297-322.

[12]   Harten, A., and P.D. Lax, "A random choice finite-difference
        scheme for hyperbolic conservation laws," SIAM J. Numer.
        Anal., 18(1981), pp. 289-315.

[13]   Harten, A., P.D. Lax and B. van Leer, "On upstream
        differencing and Godunov-type schemes for hyperbolic
        conservation laws," SIAM Rev., 25(1983), pp. 35-61.

[14]   Harten, A. and S. Osher, "Uniformly high-order accurate
        non-oscillatory schemes, I.," MRC Technical Summary Report
        #2823, May 1985, to appear in SINUM.

[15]   Harten, A., B. Engquist, S. Osher and S.R. Chakravarthy,
        "Uniformly high-order accurate non-oscillatory schemes, II.,"
        (in preparation).

[16]   Harten, A., B. Engquist, S. Osher and S.R. Chakravarthy,
        "Uniformly high-order accurate non-oscillatory schemes, III.,"
        ICASE Report No. 86-22 (April 1986).

[17]   Harten, A., S. Osher, B. Engquist and S.R. Chakravarthy,
        "Some results on uniformly high-order accurate essentially
        non-oscillatory schemes," in "Advances in Numerical and
        Applied Mathematics", Eds. J.C. South, Jr. and M.Y. Hussaini,

185

ICASE Report No. 86-18, (March 1986); also to appear in J. Appl. Num. Math.

[18] Jameson, A. and P.D. Lax, in "Advances in Numerical and Applied Mathematics", Ed. J.C. South, Jr. and M.Y. Hussaini (1986), ICASE Report No. 86-18; also to appear in J. Appl. Num. Math.

[19] P.D. Lax, "Weak solutions of nonlinear hyperbolic equations and their numerical computation," Comm. Pure Appl. Math., v. 7(1954), pp. 159-193.

[20] P.D. Lax, "Hyperbolic systems of conservation laws and the mathematical theory of shock waves," Society for Industrial and Applied Mathematics, Philadelphia, 1972.

[21] P.D. Lax and B. Wendroff, "Systems of conservation laws," Comm. Pure Appl. Math., v. 13(1960), pp. 217-237.

[22] van Leer, B., "Towards the ultimate conservative difference scheme. II. Monotonicity and conservation combined in a second order scheme," J. Computational Phys., v. 14, 1974, pp. 361-370.

[23] van Leer, B., "Towards the ultimate conservative difference schemes. V. A second order sequel to Godunov's method," J. Comp. Phys., v. 32(1979), pp. 101-136.

[24] MacCormack, R.W., "Numerical solution of the interaction of a shock wave with a laminar boundary layer (Proc. 2nd Internat. Conf. on Numerical Methods in Fluid Dynamics, M. Holt, Editor)," Lecture Notes in Phys., v. 8, Springer-Verlag, New York, 1970, pp. 151-163.

[25] Majda, A. and S. Osher, "Numerical viscosity and entropy condition," Comp. Pure Appl. Math., v. 32(1979), pp. 797-838.

[26]   Murman, E.M., "Analysis of embedded shock waves calculated
        by relaxation methods," AIAA Jour., v. 12(1974), pp. 626-633.

[27]   Osher, S. and S.R. Chakravarthy, "High-resolution schemes and
        entropy condition," SINUM, v. 21(1984), pp. 955-984.

[28]   Osher, S. and E. Tadmor, "On the convergence of difference
        approximations to conservation laws," submitted to Math.
        Comp.

[29]   Richtmyer, R.D. and K.W. Morton, "Difference methods for
        initial value problems," 2nd ed., Interscience-Wiley, New York,
        1967.

[30]   Roe, P.L., "Some contributions to the modeling of
        discontinuous flows," in Lectures in Applied Mathematics,
        v. 22(1985), pp. 163-193.

[31]   Sweby, P.K., "High resolution schemes using flux limiters for
        hyperbolic conservation laws," SINUM, v. 21(1984),
        pp. 995-1011.

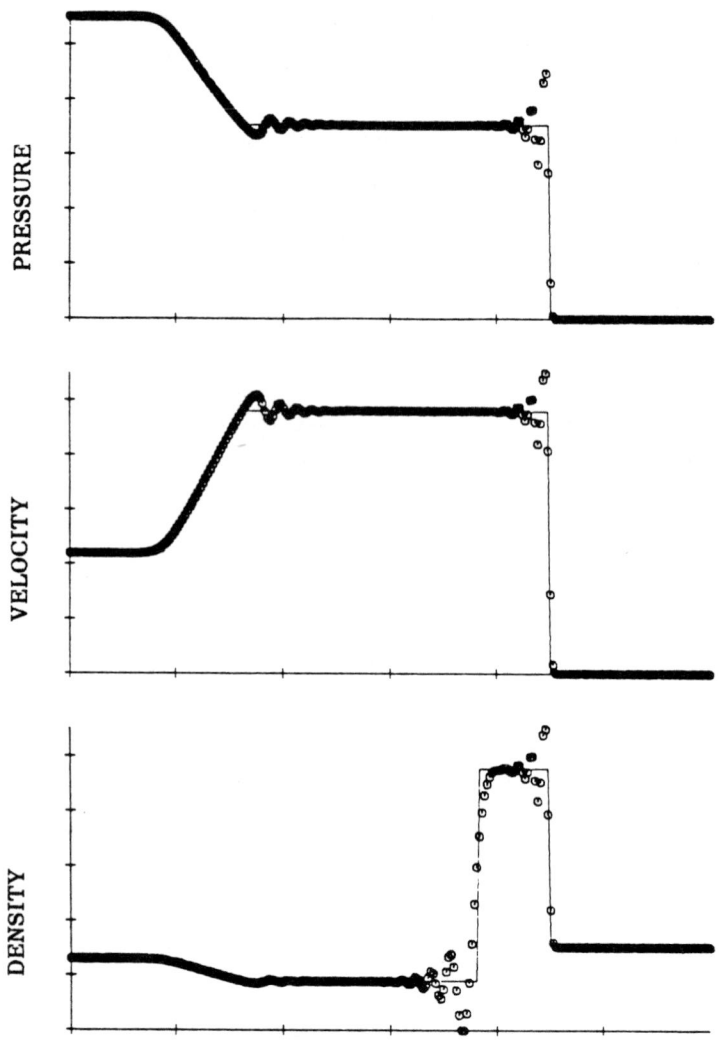

Figure 1.   MacCormack scheme.

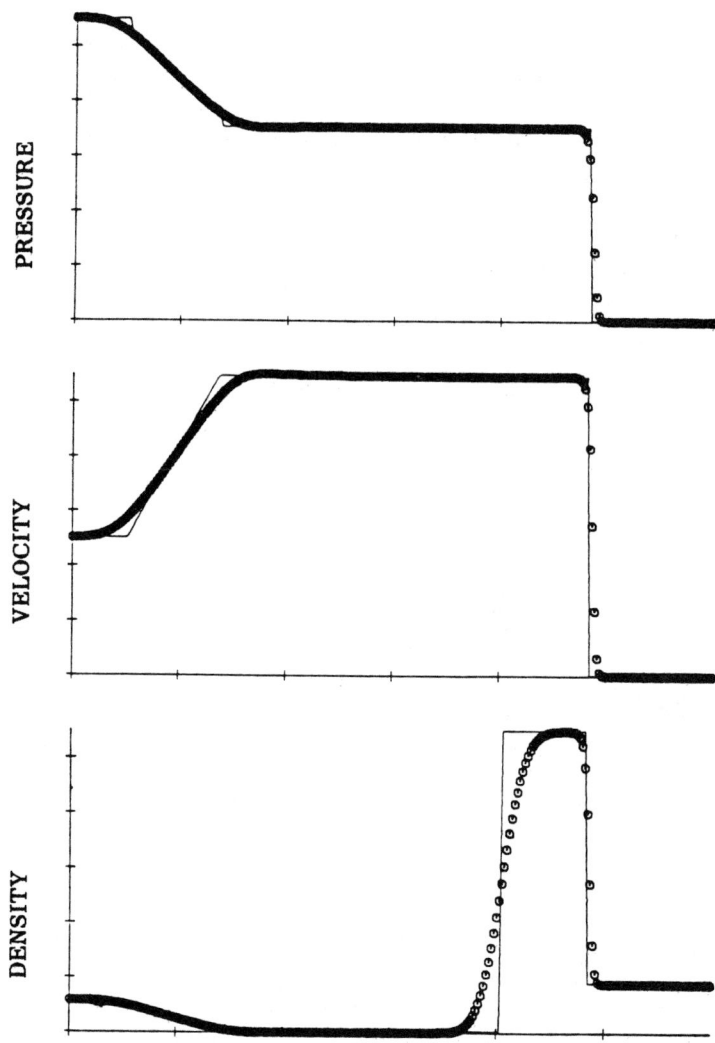

Figure 2.   The first order scheme.

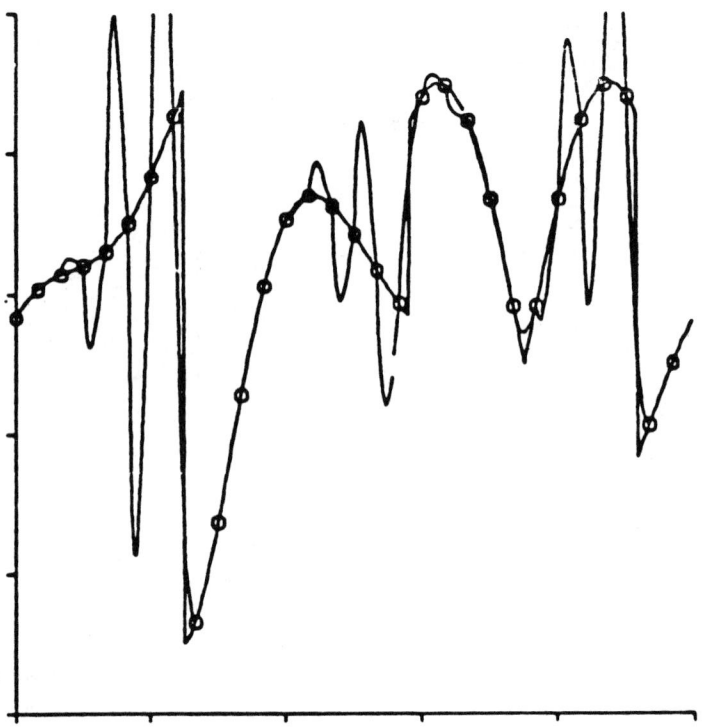

Figure 3a. $H_6(x:u)$ with a fixed stencil $i(j) = j$.

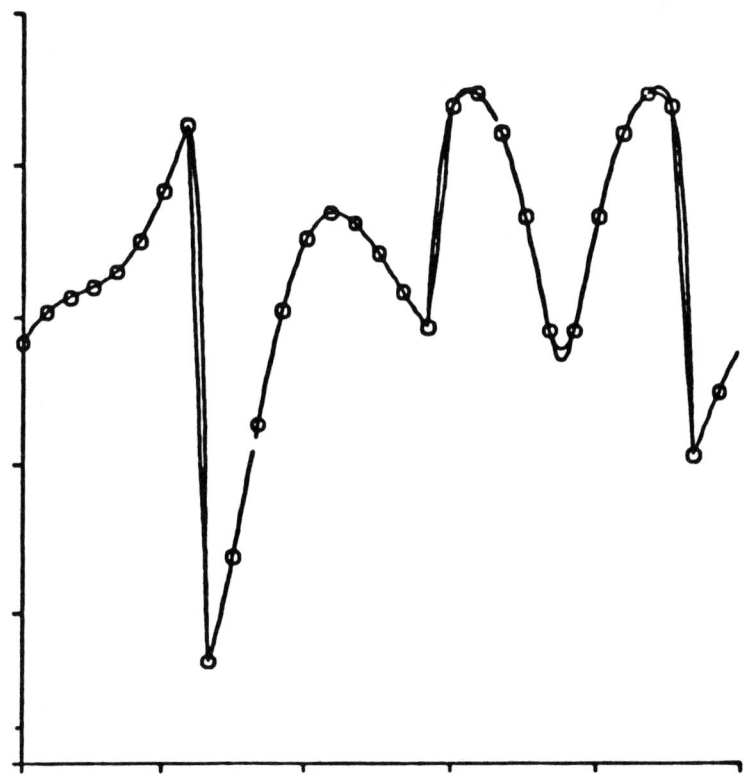

Figure 3b. $H_6(x:u)$ with an adaptive stencil (Algorithm II).

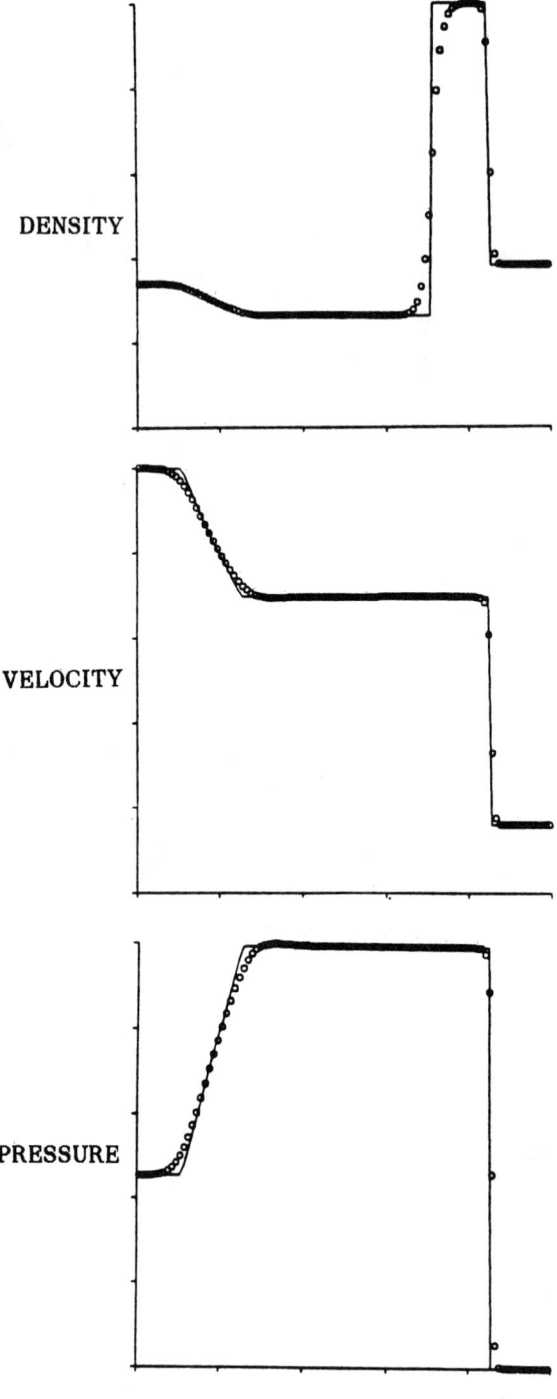

Figure 4.   Second order ENO scheme.

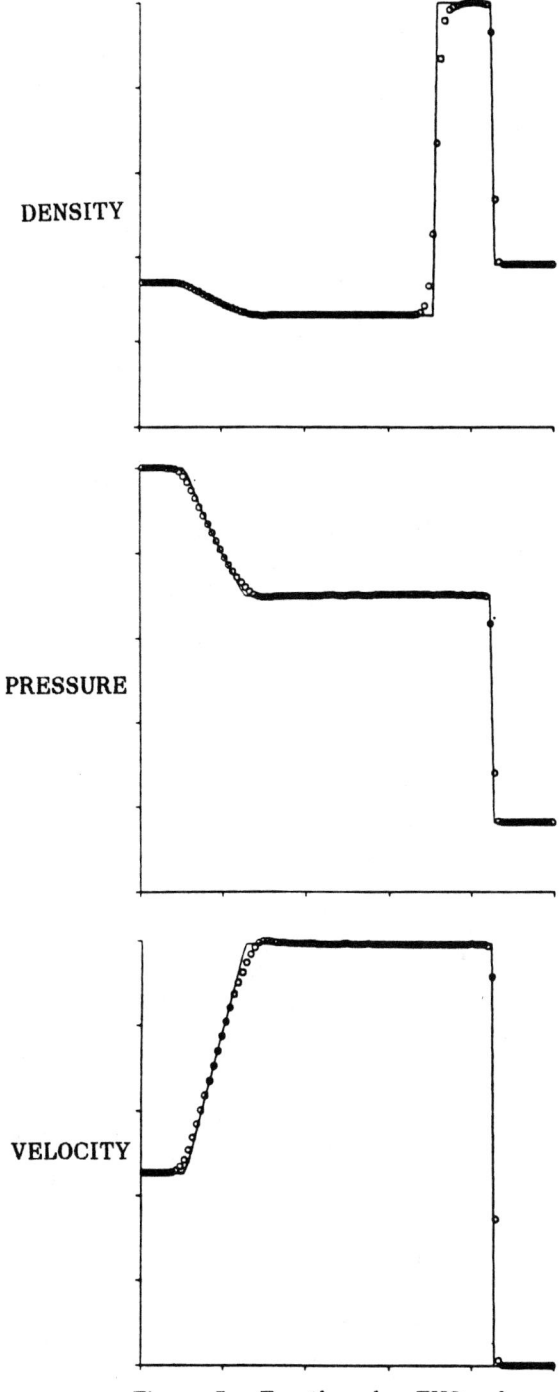

DENSITY

PRESSURE

VELOCITY

Figure 5.   Fourth order ENO scheme.

193

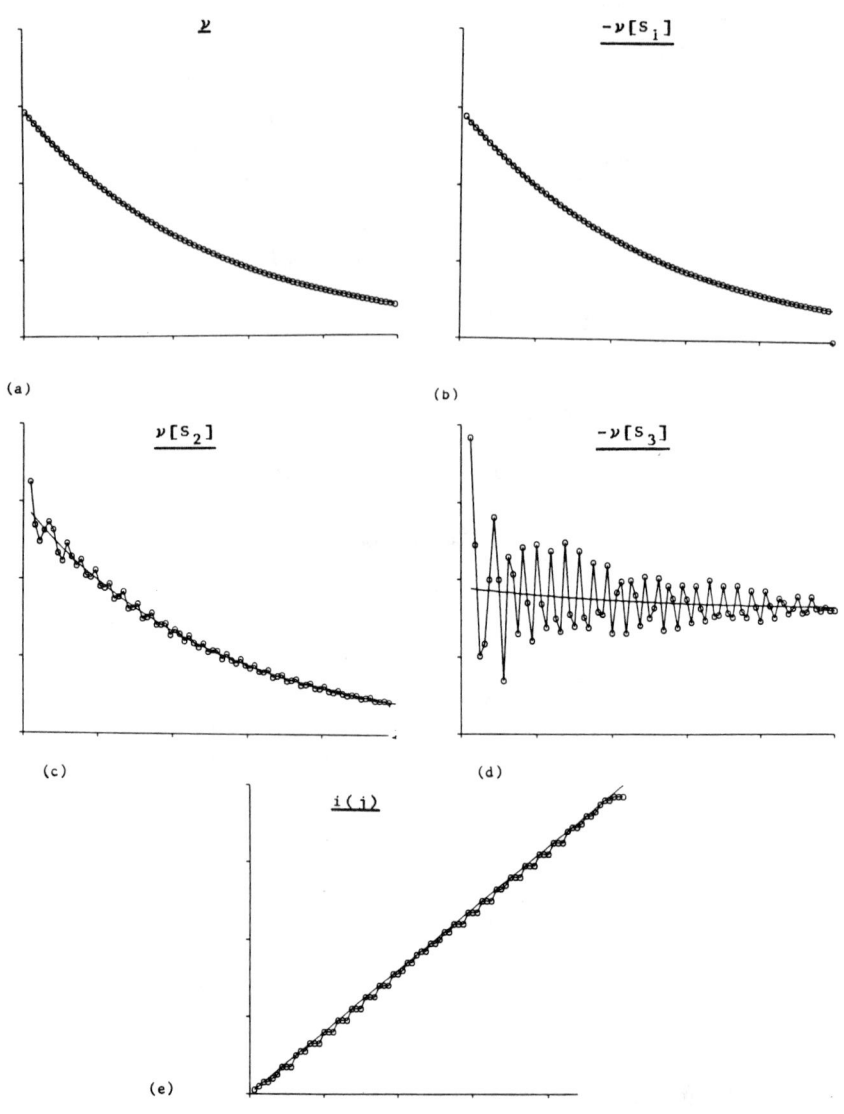

Figure 6. Solution of the IBVP (6.9) at t=1 with J=80.

194

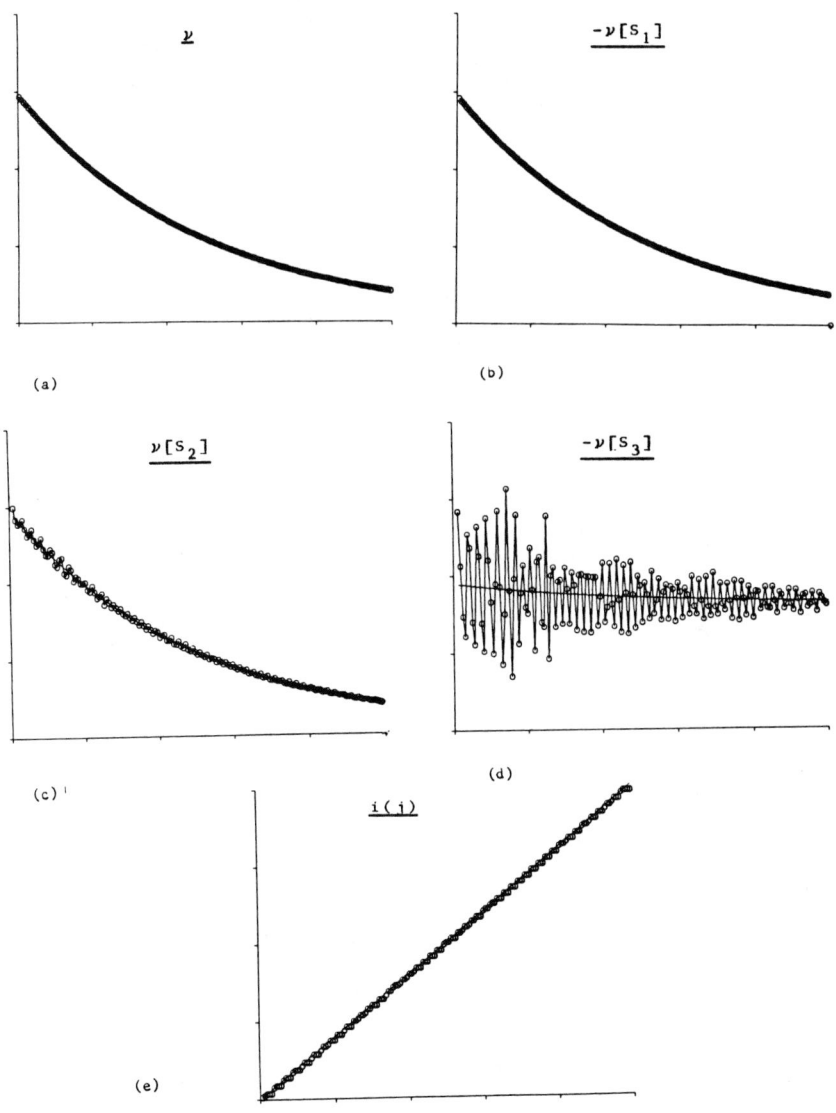

Figure 7. Solution of the IBVP (6.9) at t=1 with $\underline{J=160}$.

195

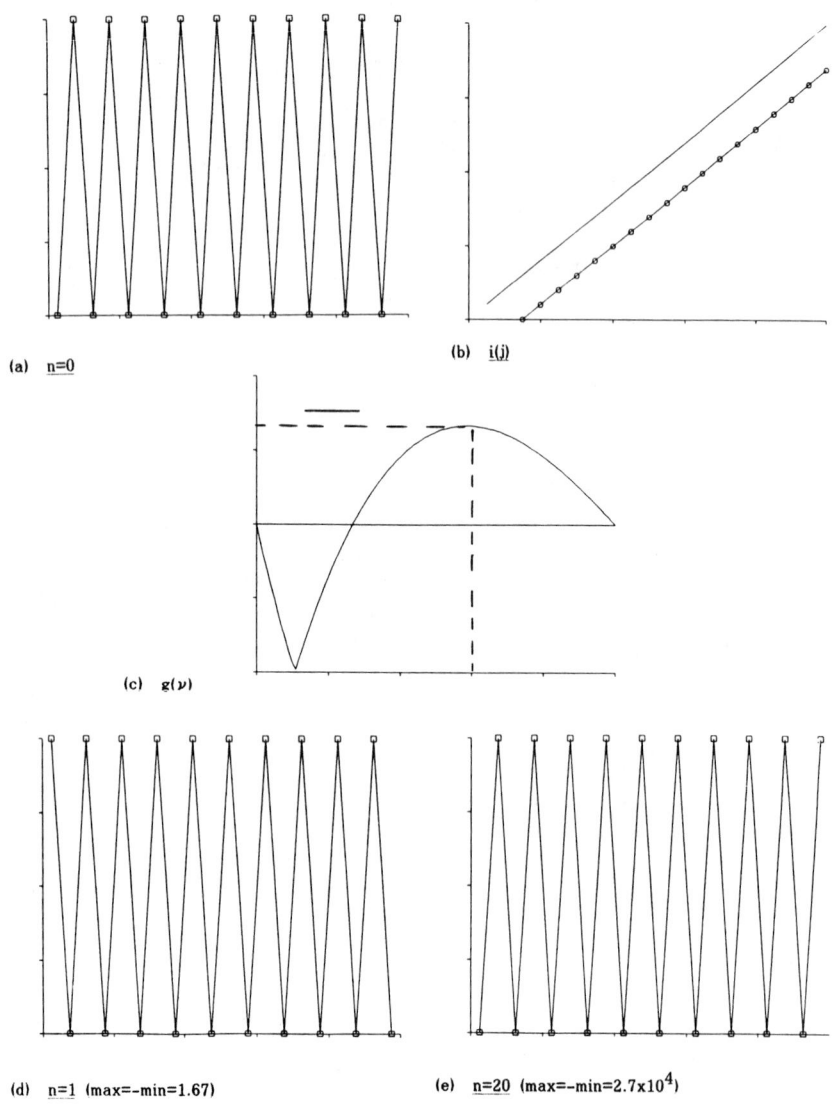

(a) n=0

(b) i(j)

(c) g(ν)

(d) n=1 (max=-min=1.67)

(e) n=20 (max=-min=2.7x10^4)

Figure 8.   Mesh-oscillation initial data.

196

n= 20   l= 0   time= 0.000E+00   dt=1
cfl= 0.600   ni= 4
eps= 0.100E-05

i	i(j)	x	$u_0(x)$
7	6	-0.950	-.9999999E+00
8	7	-0.850	0.1000000E+01
9	7	-0.750	-.9999991E+00
10	7	-0.650	0.1000000E+01
11	11	-0.550	-.1000000E+01
12	11	-0.450	0.9999993E+00
13	11	-0.350	-.9999991E+00
14	11	-0.250	0.1000001E+01
15	15	-0.150	-.1000001E+01
16	15	-0.050	0.9999993E+00
17	15	0.050	-.9999996E+00
18	15	0.150	0.1000001E+01
19	19	0.250	-.9999997E+00
20	20	0.350	0.1000001E+01
21	20	0.450	-.9999995E+00
22	20	0.550	0.9999996E+00
23	20	0.650	-.1000000E+01
24	21	0.750	0.1000001E+01
25	25	0.850	-.1000001E+01
26	26	0.950	0.9999993E+00

(a)   Initial data (numerical values)

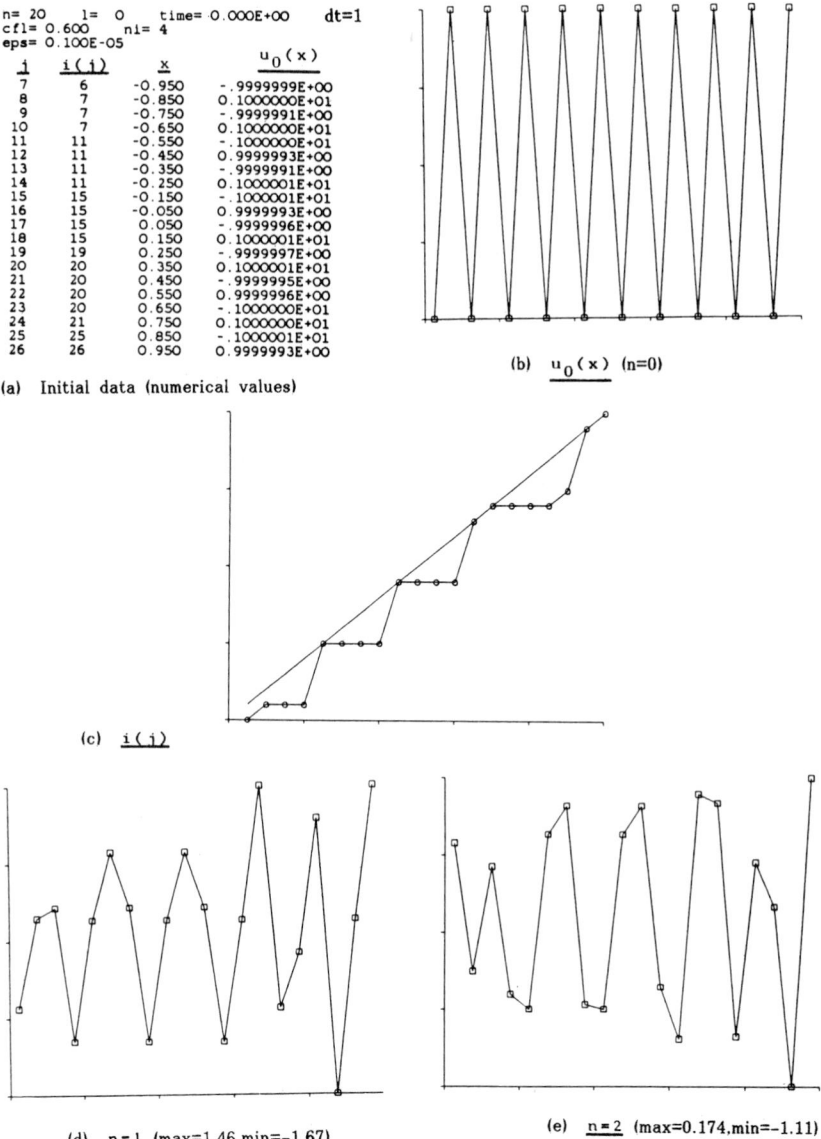

(b) $u_0(x)$ (n=0)

(c)   i(j)

(d)   n = 1   (max=1.46, min=-1.67)

(e)   n = 2   (max=0.174, min=-1.11)

Figure 9.   Randomly perturbed mesh-oscillation.

197

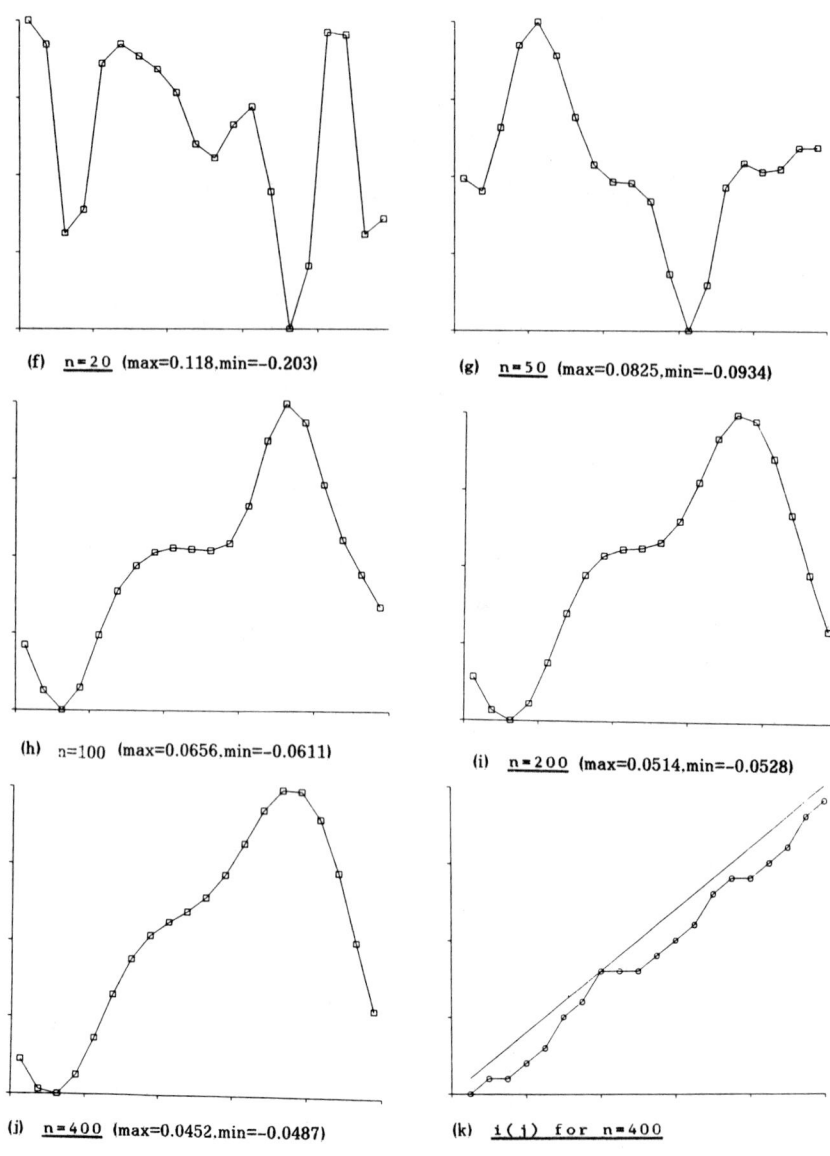

(f) n=20 (max=0.118,min=-0.203)

(g) n=50 (max=0.0825,min=-0.0934)

(h) n=100 (max=0.0656,min=-0.0611)

(i) n=200 (max=0.0514,min=-0.0528)

(j) n=400 (max=0.0452,min=-0.0487)

(k) i(j) for n=400

Figure 9. Continued

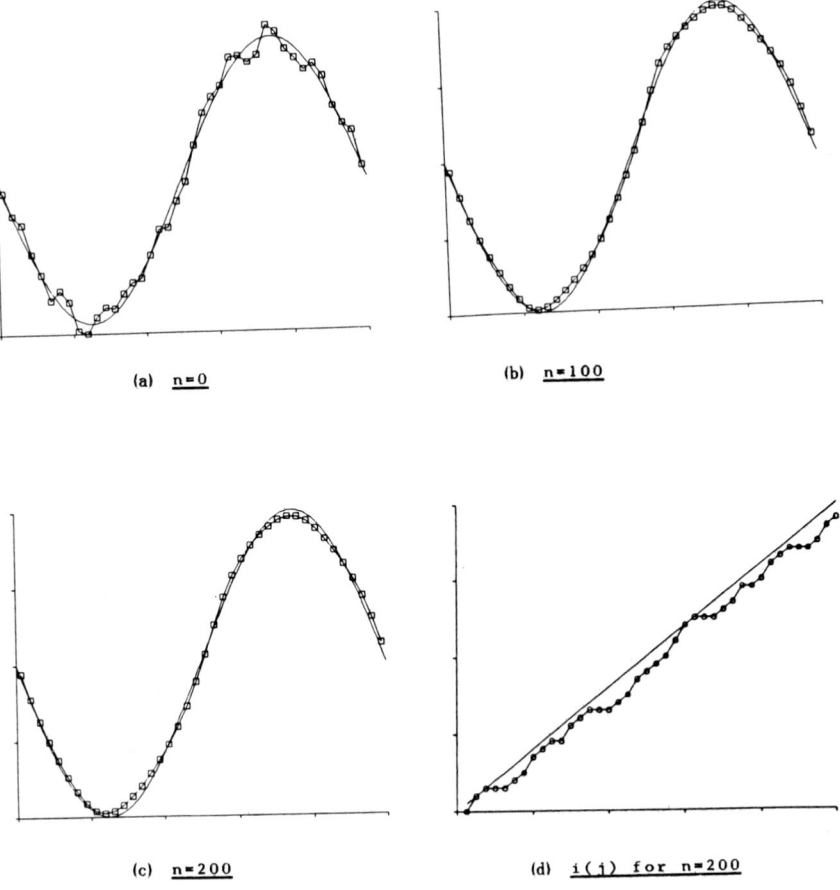

(a) n=0

(b) n=100

(c) n=200

(d) i(j) for n=200

Figure 10. Initial data of randomly perturbed sin(πx).

Table 1. Mesh-Refinement for ENO Schemes (IBVP)

$u_t + u_x = 0$.

$u_o(x) = \sin \pi x$; prescribed ata at $x = -1$, outflow at $x = 1$.

CFL = 0.8 , t = 2.

r_c is the "numerical order of accuracy."

L_∞ - Error

J	r = 2	r_c	r = 3	r_c	r = 4	r_c	r = 5	r_c	r = 6	r_c
8	1.602×10^{-1}		3.990×10^{-2}		1.846×10^{-2}		6.652×10^{-3}		8.481×10^{-3}	
		1.08		3.12		2.91		5.04		6.28
16	7.581×10^{-2}		4.593×10^{-3}		2.447×10^{-3}		2.018×10^{-4}		1.091×10^{-4}	
		1.12		2.85		2.91		4.98		5.20
32	3.488×10^{-2}		6.374×10^{-4}		3.251×10^{-4}		6.386×10^{-6}		2.972×10^{-6}	
		1.20		2.93		3.08		4.78		5.11
64	1.519×10^{-2}		8.35×10^{-5}		3.845×10^{-5}		2.312×10^{-7}		8.620×10^{-8}	

L_1 - Error

J	r = 2	r_c	r = 3	r_c	r = 4	r_c	r = 5	r_c	r = 6	r_c
8	1.374×10^{-1}		3.378×10^{-2}		1.335×10^{-2}		8.691×10^{-3}		6.632×10^{-3}	
		1.68		3.19		3.43		5.29		6.34
16	4.299×10^{-2}		3.697×10^{-3}		1.234×10^{-3}		2.227×10^{-4}		8.209×10^{-5}	
		1.67		2.84		3.66		5.13		6.39
32	1.354×10^{-2}		5.166×10^{-4}		9.742×10^{-5}		6.373×10^{-6}		9.807×10^{-7}	
		1.76		2.88		3.76		5.18		6.01
64	3.995×10^{-3}		6.994×10^{-5}		7.201×10^{-6}		1.763×10^{-7}		1.524×10^{-8}	

Table 2. Mesh Refinement for 4-th Order ENO with Exponential Data.

$$u_t + u_x = 0 \ , \quad u(x,0) = e^{-x} \ ; \ u(-1,t) = e^{1+t}; \ \text{outflow BC at } x=1.$$

$$\text{CFL} = 0.4 \ , \ t = 1.$$

J	20	40	80	160	320	640
L_∞ -error	5.063×10^{-4}	3.968×10^{-4}	4.148×10^{-4}	1.986×10^{-5}	2.648×10^{-6}	5.060×10^{-7}
L_1-error	2.905×10^{-4}	1.664×10^{-4}	9.132×10^{-5}	9.257×10^{-6}	9.648×10^{-7}	1.737×10^{-8}

HIGH FREQUENCY SIMILINEAR OSCILLATIONS

J.L. Joly
Université de Bordeaux I

J. Rauch*
University of Michigan

Dedicated to the Laxs, role models extraordinaires.

§1. Introduction

We analyze some highly oscillatory solutions of nonlinear hyperbolic partial differential equations. There is a large and very interesting literature devoted to formal high frequency solutions. We refer to [W], [K-H], [H-M-R], [F-F-McL], and the references therein for more information. Rigorous results asserting that there are exact solutions which behave like the formal ones are much less numerous, [D-M], [T1], and [J] being three from which we have learned much.

In the pioneering paper [L1], P.D. Lax showed that the basic formal expansions of linear geometric optics are asymptotic to exact solutions, and in addition, provide a powerful tool. With their aid he showed that nonhyperbolic variable coefficient operators had ill-posed Cauchy problem, and, for strictly hyperbolic operators he constructed the solution operator (modulo C^{∞}) enabling a study of the propagation of singularities. The importance of that paper in the succeeding development of the subject of partial differential equations is hard to exaggerate. Many proofs of necessary conditions follow Lax's guide and the theory of Fourier Integral Operators descends from his

*Research partially supported by the National Science Foundation under grant DMS-8601783.
Research supported in part by NSF Grant DMS-8120790.

parametrix. For nonlinear problems, the same asymptotic expansions have been used by Lax [L2] to construct families of entropies which have served well [see, [D], [R]] in implementing Tartar's program, [T2], for conservation laws.

It is our goal here to investigate nonlinear analogues of the oscillatory solutions studied in [L1]. We fix L(x,D) a k × k system of linear partial differential operators of order 1. The coefficients are supposed to be smooth functions of $x = (x_0, x_1, x_2, ..., x_d) = (x_0, x')$ and L is assumed to be strictly hyperbolic with respect to $(1,0,...,0)$. Thus, the first variable x_0 is timelike and is sometimes written t. The space dimension is d. The solutions of linear geometric optics are of the form

$$(1) \qquad v^\epsilon(x) \sim e^{i\varphi(x)/\epsilon} \sum_{j=0}^{\infty} \epsilon^j a_j(x)$$

with a_j, φ smooth, φ real valued, and $d\varphi \neq 0$. The phase φ satisfies the eikonal equation

$$(2) \qquad \det L_1(x, d\varphi(x)) = 0$$

where $L_1(x, \xi)$ is the principal symbol of L. The solutions v^ϵ are rapidly oscillating in directions transverse to the codimension one foliation $\mathcal{F}$ given by the level surfaces of φ. The leaves of $\mathcal{F}$ are characteristic, and, the construction is local, limited by the local solvability of the (nonlinear) eikonal equation.

We are interested in semilinear equations

$$(3) \qquad Lu + f(x,u) = g \in C^\infty_{(0)}([-T,T] \times \mathbb{R}^d)$$

where f is a smooth function of its arguments and

$$f(x,0) = \nabla_u f(x,0) = 0.$$

We suppose that the coefficients of L as well as their derivatives are bounded on $[-T,T] \times \mathbb{R}^d$ and $D^\alpha_{x,u} f$ is bounded on

203

$[-T,T] \times \mathbb{R}^d \times K$ for any α and compact $K \subset \mathbb{C}^k$.

One way to search for analogues of (7) is to consider f which vanish for $t < 0$ and consider (3) subject to $u|_{t<0} = v^\epsilon$, with v^ϵ linear solutions as in (1). This defines a family u^ϵ whose behavior as $\epsilon \longrightarrow 0$ is to be studied. We will call such problems, with initial data given in a **full** time band, *propagation problems* to distinguish them from the Cauchy problem.

For linear problems, the solutions v^ϵ and $\epsilon^q v^\epsilon$ are equivalent. However, for nonlinear problems they are not. Thus we will keep the factor ϵ^q which represents the relation between the amplitude and wavelength.

One can also study the Cauchy problem

$$u^\epsilon(0,x') \sim \epsilon^q \, e^{i\psi(x')/\epsilon} \sum_{j=0}^{\infty} \epsilon^j a_j(x')$$

$$a_j \in C_0^\infty(\mathbb{R}^d, \mathbb{C}^k), \ \psi \in C^\infty(\mathbb{R}^d, \mathbb{R}), \ d\psi \neq 0.$$

In this case, the linear solution is a superposition of k geometric optics solutions v_m^ϵ, $1 \leqslant m \leqslant k$ with phases φ_m satisfying the eikonal equation (2) and initial condition $\varphi_m|_{t=0} = \psi$. There are k phases, one for each solution $\tau(x)$ of $\det L_1(0,x';\tau,d_{x'}\psi(x')) = 0$. The phases φ_m define foliations $\mathcal{F}_m$ which are pairwise transverse. Near supp $u^\epsilon(0,x')$, the k waves coexist and for nonlinear problems they will interact. Thus, the Cauchy problem, in addition to the features of the propagation problem force one to consider the interaction of waves oscillating with different phases. This renders the Cauchy problem more difficult, and richer.

As a final preparatory remark we observe that the size of q is very important. Consider, for example the equation $u_t + u^2 = 0$, $u(0) = h$ with exact solution

$$u(t,x) = h(x-t)/(1 + th(x-t)).$$

With $h = \epsilon^q e^{ix/\epsilon}$ we find:

204

- For $q > 0$, the solutions exist on a band which grows with ϵ. There is an expansion in increasing powers of ϵ. The leading term, in ϵ^q, solves the linearized equation at $u = 0$. The first nonlinear correction is $0(\epsilon^{2q})$.

- For $q = 0$, the solution exists on a band independent of ϵ. Nonlinear effects occur in the principal, order ϵ^0, term. The amplitude is not small, linearization yields nothing.

- For $q < 0$, the domain of existence shrinks as $\epsilon \longrightarrow 0$, the amplitude grows, the solutions are sensitive to the behavior of the nonlinear term f as $|u| \longrightarrow \infty$.

The first two classes will be called *small oscillations and oscillations of critical size* respectively. Our positive results concern three classes of problems:

- For small oscillations in $d = 1$ we can treat the Cauchy problem while for $d > 1$ we require that the data be very small that is $q > d/2$. The range $0 < q \leqslant d/2$ is open for $d > 1$.

- For oscillations of critical size in $d = 1$ we have detailed results for the Cauchy problem including phenomena of resonance where small divisor problems are present.

- For oscillations of critical size in $d > 1$ we can treat the propagation problem with an incoming wave oscillating with one phase. We find modulated high frequency waves asymptotic to exact solutions. A complete asymptotic expansion in powers of ϵ is justified. The interaction of such waves when $k = 2$

(which excludes resonance) is also handled.

In the subsequent sections we will discuss the first and third results. The reader is referred to [J–R1] for the second and to [J–R2] for the complete proofs.

§2. **Small Oscillations**

The Cauchy problem for (1) is well posed in H^s for $s > d/2$ in the sense that if

$$\mathcal{Y}_s(T) \equiv \{g \in L^1((0,T);\ H^s(\mathbb{R}^d)):\ \forall_{1 \leqslant j \leqslant s+1},$$

$$\partial_t^j\ g \in L^1((0,T);\ H^{s-j}(\mathbb{R}^d))\}$$

$$\bar{X}_T \equiv \mathcal{Y}_s(T) \times H^s(\mathbb{R}^d)$$

$$\bar{Y}_T \equiv \bigcap_{0 \leqslant j \leqslant s+1} C^j([0,T];\ H^{s-j}(\mathbb{R}^d))$$

then for any $\{g,h\} \in \bar{X}_T$ there is a $T(g,h)\epsilon]0,\bar{T}]$ and a unique solution $u \in \bar{Y}_T$ to (1) with Cauchy data $u|_{t=0} = h$. The time of existence T can be chosen uniformly for g,h lieing in a bounded subset of $\bar{X}_T$. The map g,h $\longrightarrow$ u is C^∞ from $\bar{X}_T$ to $\bar{Y}_T$.

For $q > d/2$, choose $s \in]d/2,q[$. Then $\epsilon^q a(x')e^{i\psi(x')/\epsilon}$ tends to zero in $H^s(\mathbb{R}^d)$ so the solution can be approximated by a Taylor polynomial of degree M with error which is

$$0(\|\epsilon^q a e^{i\psi/\epsilon}\|_{H^s(\mathbb{R}^d)}^{M+1}) = 0(\epsilon^{(q-s)(M+1)}).$$

For $d = 1$ this result can be improved to $q > 0$ using the fact that the Cauchy problem is well–posed in $L^\infty(\mathbb{R}) \cap C(\mathbb{R})$. That is the solution operator $\bar{X}_T \longrightarrow \bar{Y}_T$ is equally well behaved when $d = 1$ and

$$\mathcal{Y}(T) \equiv L^1((0,T);\ L^\infty(\mathbb{R}) \cap C(\mathbb{R}))$$

206

$$\bar{X}_T \equiv \mathcal{U}(T) \times (L^\infty(\mathbb{R}) \cap C(\mathbb{R}))$$

$$\bar{Y}_T \equiv C([0,T]; L^\infty(\mathbb{R}) \cap C(\mathbb{R})).$$

For $q > 0$, $\epsilon^q a e^{i\psi/\epsilon} \longrightarrow 0$ in $L^\infty \cap C$ and a Taylor expansion is justified.

Theorem 1. Let $\underline{u}$ be the solution of $L\underline{u} + f(x,\underline{u}) = g$, $u\big|_{t=0} = 0$,

$u \in C([0,\bar{T}] \times \mathbb{R}^d)$. Suppose that $q > d/2$ or, $d = 1$ and $q > 0$. Then for ϵ small the solution u^ϵ to (1) with

$u^\epsilon\big|_{t=0} = \epsilon^q a(x') e^{i\psi(x')/\epsilon}$ exists in $[0,\bar{T}] \times \mathbb{R}^d$ and Taylor expansion

about $\underline{u}$ is justified. This yields an expansion

$$u^\epsilon \sim \sum_{\ell=0}^{\infty} \epsilon^{q\ell} P_\ell(ae^{i\psi/\epsilon}) \quad \text{with} \quad P_0 = \underline{u} \quad \text{and} \quad P_\ell: \overset{\ell}{\Pi} B \longrightarrow \bar{X}_T$$

ℓ-linear with $B = L^\infty \cap C$ if $d = 1$ or $B = H^s$. The expansion is asymptotic in the sense that

$$\sup_{[0,T] \times \mathbb{R}^d} \left| (u^\epsilon - \sum_{\ell \leq N-1} \epsilon^{q\ell} P_\ell) \right| \leq C_N \epsilon^{N\ell}$$

For the small oscillations treated in this Theorem one can describe simply and in detail some important nonlinear phenomena including the generation of harmonics and resonance. Toward this end, we remark that an efficient way to compute the Taylor series in the theorem is to add an artificial parameter α and consider the solutions v^α with initial data equal to α times the Cauchy data of u^ϵ. A formal power series $v^\alpha \sim \sum_{n=0}^{\infty} \alpha^\ell v_\ell$ is found, and finally α is set equal to one. The first corrector $\epsilon^q P_1$ is determined by

$$LP_1 + f_u(x,\underline{u})P_1 = 0, \quad u_1\big|_{t=0} = ae^{i\psi/\epsilon}.$$

207

Thus P_1 is given, locally in time, by a geometric optics expansion

$$\epsilon^q P_1 = \sum_{j=1}^{k} \epsilon^q a_m(x) e^{i\varphi_m(x)/\epsilon} + 0(\epsilon^{q+1})$$

To determine the succeeding corrections $\epsilon^{q\ell} P_\ell$ one solves linear equations

$$L P_\ell + f_u(x,\underline{u}) P_\ell = N$$

when N is a nonlinear function of $P_0, P_1, ..., P_{\ell-1}$. In particular, one finds source terms which are nonlinear functions of $e^{i\varphi_1/\epsilon}, e^{i\varphi_2/\epsilon}, ..., e^{i\varphi_m/\epsilon}$. Considering the complex conjugate and products yields monomials of the form

$$\epsilon^{(|\alpha|+|\beta|)q} a(x) \exp i(\sum_m \alpha_m \varphi_m - \beta_m \varphi_m), \qquad \alpha, \beta \in \mathbb{N}^k$$

and one is lead to consider the response

(4)
$$(L + f_u)^{-1} \{ a(x) \exp i(\sum(\alpha_m - \beta_m)\varphi_m) \}.$$

We mention four distinct possibilities each representing a different type of nonlinear interaction. We begin with interactions involving only one phase. Thus suppose $\alpha_m = \beta_m = 0$ unless $m = j$. The term (4) is then of the form

(5)
$$b(x) e^{i(\alpha_j - \beta_j)\varphi_j/\epsilon} + 0(\epsilon).$$

(a)   **Cancellation.** If $\alpha_j - \beta_j = 0$ one finds a nonoscillatory response. This should not be confused with the linear cancellation of oscillation which can occur by interference on *adding* to an oscillatory wave its negative. Here we find a nonoscillatory term on *multiplying* an oscillatory term by its complex conjugate.

208

(b)   **Excitation of overtones**.   If $\alpha_j - \beta_j \neq 0$, (5) is an

overtone of the fundamental $e^{i\varphi_j/\epsilon}$.

Next we turn to the interaction of oscillations with different phases.   When only one phase is involved the resulting phase $(\alpha_j - \beta_j)\varphi_j$ always satisfies the eikonal equation.

(c)   **Nonresonant interaction**.   If for all x,

(6) $$\det L_1(x, d\Sigma(\alpha_m - \beta_m)\varphi_m(x)) \neq 0,$$

the inversion (4) is essentially elliptic.   One finds (see for example [H, page 272])

$$\epsilon b(x)e^{i\Sigma(\alpha_m - \beta_m)\varphi/\epsilon} + 0(\epsilon^2),$$

$$b(x) = L_1(x, d\Sigma(\alpha_m - \beta_m)\varphi_m(x))^{-1}a(x).$$

Thus, the response is smaller by a factor of ϵ, than one might guess and is local, there is no propagation of these nonresonant terms.

(d)   **Resonant interaction**.   If $d(\Sigma(\alpha_m - \beta_m)\varphi) \neq 0$, and

(7) $$\det L_1(x, d\Sigma(\alpha_m - \beta_m)\varphi_m(x)) \equiv 0$$

the inversion (4) is computed by geometrical optics as,

$$b(x)e^{i\Sigma(\alpha_m - \beta_m)\varphi/\epsilon} + 0(\epsilon)$$

the $b(x)$ determined by integrating ordinary differential equations along the rays corresponding to the phase $\Sigma(\alpha_m - \beta_m)\varphi_m$.   Thus we have created an oscillation with

209

a new phase, $\sum(\alpha_m - \beta_m)\varphi_m$. This new wave usually propagates beyond the support of $a(x)$ according to standard transport equations. Naturally, there is an enormous number of possibilities between the extreme cases (6), and (7). These intermediate cases can be quite complicated.

The reader is referred to [J-R2] for a discussion of more phenomena including a sum law estimating the magnitude of corrections due to intermode interactions and an example of chaotic interaction of three modes for a semilinear wave equation (predicted in [H-M-R]).

§3. Oscillations of Critical Size

The propagation and *interaction* of critical size oscillations in one dimension are discussed in [J-R1], [J-R2]. Here we discuss the propagation problem for one mode oscillations with d arbitrary. The main result is a justification of a modulated wave approximation

(8) $$u^\epsilon = A(x, \varphi(x)/\epsilon) + 0(\epsilon).$$

Starting from a purely sinusoidal oscillation, a $e^{i\varphi/\epsilon}$, we expect to find, from nonlinear effects, terms of the form

$$L^{-1}(\psi(x)e^{in\varphi/\epsilon}), \qquad n \in \mathbf{Z}.$$

One cannot expect anything simpler than

$$\sum_n a_n(x)e^{in\varphi/\epsilon} = A(x, \varphi/\epsilon),$$

$$A(x, \theta) \equiv \sum_{n=-\infty}^{\infty} a_n(x)e^{in\theta}.$$

Thus, one is lead inexorably to such modulated waves with profile A. Clearly, superposition of geometric optics solutions with phases $n\varphi$, $n \in \mathbf{Z}$ yield linear waves with such profiles. We find nonlinear solutions of form (8) and, more generally, complete asymptotic expansions

$$u^\epsilon \sim \sum_{m=0}^{\infty} \epsilon^m A_m(x, \varphi(x)/\epsilon).$$

We consider the propagation problem: given u^ϵ in $t < 0$ of form (8) in $t < 0$, show that a solution u^ϵ exists on an interval $t < T$, with $T > 0$ independent of ϵ and continues to have form (8). One also seeks an algorithm for computing A, and more generally A_m, from their values in the past.

Considering an equation with $f \equiv 0$ in $t < 0$ and $u^\epsilon|_{t<0}$ a linear wave of Lax, provides problems of this type. Thus, once the above result is demonstrated we conclude that nonlinear waves of form (8) exist and propagate maintaining that form.

Fix Γ a *bounded* open subset of $]t_1, t_2[\times \mathbb{R}^d$ with $t_1 < 0 < t_2$, $\varphi \in C^\infty([t_1, t_2] \times \mathbb{R}^d, \mathbb{R})$, $d\varphi \neq 0$ on $\bar{\Gamma}$, satisfying the eikonal equation (2) on a neighborhood of $\bar{\Gamma}$. Suppose, in addition, that backward null bicharacteristics starting over $\bar{\Gamma}$ remain over $\bar{\Gamma}$ for $t \geq t_1$. Let $\mathcal{F}$ be the foliation of Γ by level surfaces of φ, and let $\mathcal{V}$ be the Lie algebra of smooth vector fields defined on a neighborhood of $\bar{\Gamma}$ and tangent to $\mathcal{F}$. We will work in the Fréchet spaces H_T, $T \in]t_1, t_2]$, defined by

$$\Gamma_T \equiv \Gamma \cap \{t \leq T\}$$

$$H_T \equiv \{u \in L^\infty(\Gamma_T): N \in \mathbb{N}, V_1, ..., V_N \in \mathcal{V} \Rightarrow$$

$$V_1 V_2 ... V_N u \in L^2(\Gamma)\}.$$

Theorem 2 [R-R]. Given v^ϵ bounded in H_0, g^ϵ bounded in H_T with

$$Lv^\epsilon + f(x, v^\epsilon) = g^\epsilon \text{ in } \Gamma_0$$

then there is a $\bar{T} \in]0, T]$ and unique u^ϵ bounded in $H_{\bar{T}}$ solutions to

$$Lu^\epsilon + f(x, u^\epsilon) = g^\epsilon \text{ in } \Gamma_{\bar{T}}, \quad u^\epsilon|_{\Gamma_0} = v^\epsilon.$$

211

Thus, if

(9)
$$v^\epsilon = A_0(x,\varphi(x)/\epsilon) + o(1) \text{ in } \Gamma_0$$

(10)
$$g^\epsilon = G(x,\varphi(x)/\epsilon) + o(1) \text{ in } \Gamma_T$$

when o(1) is measured in H_0 and H_T respectively, the theorem shows that the v^ϵ continue to a fixed interval $t < \bar{T}$ and are bounded in $H_{\bar{T}}$. This existence is nontrivial since in H^s_{loc} with $s > 0$ the size of v^ϵ diverges to infinity as ϵ decreases to zero. Thus, the standard bounds for the time of breakdown are o(1) as ϵ converges to zero.

Theorem 3 [J-R2]. (1) If (9), (10) hold then there is a unique $A \in C^\infty(\bar{\Gamma}_T \times S^1)$ such that (8) holds in $\Gamma_{\bar{T}}$, and o(1) measured in $H_{\bar{T}}$. (2) In coordinates so that $\varphi = x_1$, $L = \partial_0 + M_1(x)\partial_1 + ...$, $M_1 = \text{diag}(0,\lambda_2(x),...,\lambda_k(x))$, the profile A is uniquely determined by the integro-differential system:

(11)
$$\begin{cases} A(x,\theta) = (a_1(x,\theta),a_2(x),...,a_k(x)) \\ A|_{t<0} = A_0 \\ \mathcal{E}(L(x,D_x)A(x,\theta) + f(x,A(x,\theta)) - G(x,\theta)) = 0 \end{cases}$$

where

$$\mathcal{E} \equiv \text{diag}(I,E,...,E)$$

$$E\ell(x,\theta) \equiv (2\pi)^{-1} \int_0^{2\pi} \ell(x,\theta)d\theta.$$

Example. To see clearly the form of the equations determining A we present a simple example in detail.

$$L = \partial_0 + \begin{bmatrix} 0 & 0 \\ 0 & 1 \end{bmatrix} \partial_1, \quad f = (f_1(u),f_2(u)), \quad \varphi = x_1.$$

The profile is $(a_1(x,\theta),a_2(x))$. The equations for a_i are

212

$$(12) \quad \partial_0 a_1(x,\theta) + f_1(a_1(x,\theta), a_2(x)) - g_1(x,\theta) = 0$$
$$(\partial_0 + \partial_1) a_2(x) + E\{f_2(a_1(x,\theta), a_2(x)) - g_2(x,\theta)\} = 0$$

with a_1, a_2 determined in $t < 0$ from data. Note that the speed one waves, described by a_2, are influenced only by mean values of g and $f(A)$, while the speed zero waves propagate the detailed structure. As the speed one waves cross the oscillations transversally, this averaging is not surprising. Note that the mean values of A in $t < 0$ do not determine the mean values of A in $t > 0$. This is a typically nonlinear effect.

Outline of the Proof of Theorem 3. First one shows that the equations (11) have a unique solution A. Then, for the difference

$$\delta^\epsilon(x) \equiv u^\epsilon(x) - A(x, \varphi(x)/\epsilon)$$

one derives an equation of the form

$$L\delta^\epsilon + M(x)\delta^\epsilon = \beta^\epsilon(x)$$

where

$$\beta^\epsilon(x) = B(x, \varphi(x)/\epsilon), \qquad \int_0^{2\pi} B(x,\theta) d\theta = 0.$$

Thus β^ϵ tends *weakly* to zero. The energy method is used with the goal of showing that δ^ϵ tends *strongly* to zero. For symmetric hyperbolic L (the nonsymmetric case is somewhat trickier), we multiply by δ^ϵ and must show that

$$(13) \qquad \int_0^t \int <\delta^\epsilon(x), \beta^\epsilon(x)> \, dx \, dt = o(1).$$

From Theorem 2 we know that δ^ϵ is bounded in L^2, which is not sufficient. In the coordinates of part 2 of Theorem 3 we show that the first component of β, β_1^ϵ, tends strongly to zero and that the last components of δ^ϵ, δ_m^ϵ, $m \geqslant 2$, are bounded in H^1. Together these imply (13).

Details, analytic consequences, complete asymptotic expansions, the nonsymmetric case, and a discussion of interacting oscillatory wave trains can be found in [J-R2].

References

[D]   R. DiPerna, Convergence of the viscosity method for isentropic gas dynamics, Comm. Math. Phys. 91 (1983) 1–30.

[D-M]   R. DiPerna and A. Majda, The validity of geometrical optics for weak solutions of conservation laws, Comm. Math. Phys. 98 (1985) 313–347.

[F-F-McL] H. Flaschka, G. Forest, D. McLaughlin, Multiphase averaging and the inverse spectral solution of the Korteweg-de Vries equation, Comm. Pure Appl. Math. 33 (1980) 739–784.

[H]   L. Hormander, The Analysis of Linear Partial Differential Operators I, Springer-Verlag, Berlin-Heidelberg, 1983.

[J]   J.L. Joly, Sur la propagation des oscillations par un systeme hyperbolique semi-linéaire en dimension 1 d'espace, C. R. Acad. Sc. Paris, t. 296 (25 Avril 1983).

[J-R1]   J.L. Joly and J. Rauch, Ondes oscillantes semi-linéaire en dimension 1, Proceedings Journeés E.D.P. St. Jean de Monts, 1986.

[J-R2]   J.L. Joly and J. Rauch, in preparation.

[K-H]   J. Keller and J. Hunter, Weakly nonlinear high frequency waves, Comm. Pure Appl. Math. XXXVI (1983) 547–569.

215

[H-M-R]   J.K. Hunter, A. Majda, and R. Rosales, Resonantly interacting, weakly nonlinear hyperbolic waves II: several space variables, preprint.

[L1]   P.D. Lax, Asymptotic solutions of oscillatory initial value problems, Duke Math. J. 24 (1957) 627–646.

[L2]   P.D. Lax, Shock waves and entropy, in Contributions to Nonlinear Functional Analysis, ed. E.A. Zarantonello, Academic Press (1971).

[R]   M. Rascle, Un resultat de "compacité par compensation à coefficients variable". application à l'elasticité nonlinéaire. C. R. Acad. Sc. Paris, 1986.

[R-R]   J. Rauch and M. Reed, Bounded stratified and striated solutions of hyperbolic systems, in Nonlinear Partial Differential Equation and Their Applications V, College de France Seminar H. Brezis et J.L. Lions eds. Pittman Publishers.

[T1]   L. Tartar, Solutions oscillantes des équations de Carlman, Seminaire Goulaouic-Meyer-Schwartz, (1983).

[T2]   L. Tartar, The compensated compactness method applied to systems of conservation laws, in Systems of Nonlinear Partial Differential Equations, J. Ball ed. Nato ASI Series, C. Reidel Publ. (1983).

[W]   G.B. Whitham, Linear and Nonlinear Waves, Wiley, New York, 1974.

EXACT CONTROLLABILITY AND SINGULAR PERTURBATIONS

By

J.L. Lions
College de France and CNES

Dedicated to P. Lax

1. Introduction

1.1   Let Ω be a bounded open set of $\mathbb{R}^n$, with smooth boundary Γ.

In the cylinder $\Omega \times]0,T[$ we consider the *evolution equation*

$$(1.1) \qquad \frac{\partial^2 y}{\partial t^2} + Ay = 0;$$

in (1.1) A is a linear elliptic operator, of order say 2m, and which is symmetric.

Initial conditions are given:

$$(1.2) \qquad y(0) = y^0, \quad y'(0) = y^1 \quad \text{in } \Omega$$

where we have set

$$y(0) = y(\cdot,0)$$

and

$$y'(0) = \frac{\partial}{\partial t} y(\cdot,0).$$

One can act on the system through *boundary conditions*. Let $\{B_j \mid 1 \leqslant j \leqslant m\}$ a set of m *boundary* operators, such that $\{A,B_j\}$ defines a well set boundary value problem,

---

Research supported in part by NSF Grant DMS-8120790.

supposed to be symmetric.

We consider then the non-homogeneous boundary value problem: (1.1) (1.2) and

(1.3) $B_j y = v_j$ on Σ, $1 \leqslant j \leqslant m$.

In (1.3) the v_j's are *control functions*. The solution of (1.1) (1.2) (1.3) (defined in an appropriate weak sense) is the *state y of the system*. One says that the system is *exactly controllable* if, *given* T, one can find controls v_j such that

(1.4) $y(T) = y'(T) = 0$.

In other words, one wants to drive the system to rest at time T.

Remark 1.1: In general (but not always) one will have to assume T *large enough*. This condition is clearly necessary for *hyperbolic* systems. □

Remark 1.2: One can consider more general questions by *adding constraints* to the v_j's--always with the goal of achieving (1.4). These constraints can be of the following form

(i) $v_j = 0$ except for a *subset* of indices $1,...,m$;

(ii) v_j has support in a *subset* Σ_0 of Σ (i.e. one can act only on *part* of the boundary).

One can of course also have the two constraints at the same time: this will be the case below. □

The "generic" result for exact controllability is--in a very fuzzy form!--that provided T is chosen large enough, *and provided* y^0, y^1 *are chosen in suitable Hilbert spaces*, then the system is exactly controllable.

218

The key point--as proven in J.L. Lions [1] (cf. [2] [3] for details) is *to define a Hilbert structure on the set of initial datas y^0, y^1 associated with a uniqueness result.* [We shall recall this point in Section 4 below].

The method given in this way is *constructive*, and it defines the controls v_j which drive the system to rest at time T *and* which minimize some norms on the v_j's.

1.2 Exact controllability and perturbations

Let us now consider instead of an operator **A** a *family of operators* A_ϵ:

$$(1.5) \qquad \frac{\partial^2 y_\epsilon}{\partial t^2} + A_\epsilon y_\epsilon = 0,$$

subject to (1.2) and to

$$(1.6) \qquad B_{j\epsilon} y_\epsilon = v_j \qquad \text{on } \Sigma, \ 1 \leqslant j \leqslant m$$

(or for only *some* j's, with $v_j = 0$ for the other j's; or with v_j with support in $\Sigma_0 \subset \Sigma$).

We shall say that the system (1.5) is *uniformly* exactly controllable if one can find T_0 *independent of* ϵ such that for $T > T_0$ the system (1.5) is exactly controllable; in other words, one can find $v_{j\epsilon}$ which drive the system to rest at time T.

The next (and main) step is to study the behavior of the "optimal" $v_{j\epsilon}$'s as $\epsilon \longrightarrow 0$. □

One can study this "general program" in the following situations:

(j) A_ϵ is a *singular perturbation* of A_0;

(jj)   the A_ϵ's are operators with rapidly varying coefficients, or with stochastic coefficients: is it then possible, and in which way, to "replace" A_ϵ by its "homogenized" operator?

219

(jjj)   the operators A_ϵ correspond to "single" operators defined in "singular" geometrical domains (thin plates, perforated domains).

We want to present here *an example* of a problem of type (j).

1.3   Example
We consider the equation

(1.7) $$y_\epsilon'' + \epsilon\Delta^2 y_\epsilon - \Delta y_\epsilon = 0 \quad \text{in } \Omega \times]0,T[,$$

with the initial conditions

(1.8) $$y_\epsilon(0) = y^0, \quad y_\epsilon'(0) = y^1.$$

We are going to act on the system only through $\dfrac{\partial y_\epsilon}{\partial v}$ *on part of the boundary.*

Let x^0 be given *arbitrarily* in $\mathbb{R}^n$. We set

(1.9) $$m(x) = \{m_k(x)\} = \{x_k - x_k^0\}$$

(1.10) $\begin{vmatrix} \Gamma(x^0) = \{x \mid x \in \Gamma = \partial\Omega, \ m(x)v(x) \geqslant 0\} \\ \text{where } v(x) = \text{unit normal to } \Gamma \text{ at } x \text{ directed} \\ \text{toward the exterior of } \Omega, \end{vmatrix}$

(1.11) $$\Gamma^*(x^0) = \Gamma \setminus \Gamma(x^0).$$

Then we define the *state* y_ϵ of the system as the solution of (1.7) (1.8) and

(1.12) $\begin{vmatrix} \qquad\qquad y_\epsilon = 0 \text{ on } \Sigma = \Gamma \times (0,T), \\ \\ \dfrac{\partial y_\epsilon}{\partial v} = \begin{vmatrix} v & \text{on } \Sigma(x^0) = \Gamma(x^0) \times (0,T) \\ 0 & \text{on } \Sigma^*(x^0) = \Gamma^*(x^0) \times (0,T). \end{vmatrix} \end{vmatrix}$

We use the Sobolev spaces $H_0^1(\Omega), H_0^2(\Omega)\ldots$ and their dual spaces $H^{-1}(\Omega), H^{-2}(\Omega)\ldots$.

All function spaces considered here are *real*.

The first main result that we want to prove in this paper is that (1.7) (1.8) (1.12) is *uniformly* exactly controllable.

More precisely:

Theorem 1.1: *Let us assume that*

$$(1.13) \qquad y^0 \in L^2(\Omega), \quad y^1 \in H^{-1}(\Omega).$$

Then there exists T_0, independent of ϵ, such that if $T > T_0$, there exists $v_\epsilon \in L^2(\Sigma(x^0))$ which drives the system to rest at time T.

Remark 1.3: In the course of the proof, we shall obtain *estimates* on T_0. □

Remark 1.4: The method used to prove Theorem 1.1 naturally leads to the control v_ϵ which, among all v's which drive the system to rest at time T, *minimizes* $\int_{\Sigma(x^0)} v^2 d\Sigma$. □

The main result of the paper is to study the behavior of v_ϵ as $\epsilon \longrightarrow 0$. We shall show:

Theorem 1.2: *We assume that (1.13) holds true and that T is given $> T_0$. Let v_ϵ be the control minimizing*

$\int_{\Sigma(x^0)} v^2 d\Sigma$. *Then, as $\epsilon \longrightarrow 0$, one has*

$$(1.14) \qquad -\sqrt{\epsilon}\,v_\epsilon \longrightarrow v_0 \ in \ L^2(\Sigma(x^0)) \ weakly,$$

$$(1.15) \qquad y_\epsilon \longrightarrow y_0 \ in \ C([0,T];L^2(\Omega)) \ weak \ star,$$

221

where

$$(1.16) \quad \begin{vmatrix} y_0'' - \Delta y_0 = 0, \\ \\ y_0(0) = y^0, \quad y_0'(0) = y^1 \text{ in } \Omega, \\ \\ y_0 = \begin{vmatrix} v_0 & \text{on } \Sigma(x^0) \\ 0 & \text{on } \Sigma^*(x^0) \end{vmatrix} \end{vmatrix}$$

and

$$(1.17) \quad y_0(T) = y_0'(T) = 0,$$

$$(1.18) \quad \begin{vmatrix} v_0 \text{ } minimizes \quad \int_{\Sigma(x^0)} v^2 d\Sigma \text{ } among \text{ } all \text{ } the \\ \\ controls \text{ } v \in L^2(\Sigma(x^0)) \text{ } which \text{ } drive \\ the \text{ } wave \text{ } equation \text{ } to \text{ } rest. \quad \square \end{vmatrix}$$

Remark 1.5: We have therefore here a "switching" of the boundary conditions: $y_\epsilon = 0$ on Σ and one can act on $\frac{\partial y_\epsilon}{\partial v}$ on part of Σ, namely $\Sigma(x^0)$. For the limit system, one acts on y (on the set $\Sigma(x^0)$), and not on $\frac{\partial y}{\partial v}$. $\square$

1.4 Organization of the paper

The proof of Theorem 1.2 will rely on the general method introduced in J.L. Lions [1] for exact controllability and on results of singular perturbations which are independent of control theory.

In Section 2, we study the equation

$$(1.19) \quad \begin{vmatrix} u_\epsilon'' + \epsilon \Delta^2 u_\epsilon - \Delta u_\epsilon = f_\epsilon, \\ \\ u_\epsilon(0) = u^0, \quad u_\epsilon'(0) = u^1, \\ \\ u_\epsilon = \frac{\partial u_\epsilon}{\partial v} = 0 \text{ on } \Sigma \end{vmatrix}$$

where

(1.20) $$f_\epsilon \longrightarrow f \text{ in } L^1(0,T;L^2(\Omega))$$

and we study the behavior of $\sqrt{\epsilon}\Delta u_\epsilon|_\Sigma$ as $\epsilon \longrightarrow 0$.

In Section 3 we consider a related problem for non-homogeneous problems (Section 3 is, in a sense, the "dual" of Section 2).

In Section 4 we explain, in the particular example at hand, the general method for exact controllability (and which leads to apparently new uniqueness results).

The proof of the main results is then given in Section 5.

2. **Singular perturbations. Strong solutions**

    2.1 **Setting of the problem**

    Let f_ϵ be a sequence of functions in $L^1(0,T;L^2(\Omega))$ such that

(2.1) $$f_\epsilon \longrightarrow f \text{ in } L^1(0,T;L^2(\Omega)) \text{ as } \epsilon \longrightarrow 0.$$

Theorem 2.1: *Let u_ϵ be the solution of*

(2.2) $$u_\epsilon'' + \epsilon\Delta^2 u_\epsilon - \Delta u_\epsilon = f_\epsilon \text{ in } \Omega \times]0,T[$$

such that

(2.3) $$u_\epsilon(0) = u_\epsilon^0, \quad u_\epsilon'(0) = u_\epsilon^1$$

where

(2.4) $$\{u_\epsilon^0, u_\epsilon^1\} \longrightarrow \{u^0, u^1\} \text{ in } H_0^1(\Omega) \times L^2(\Omega)$$

(2.5) $$u_\epsilon^0 \in H_0^2(\Omega), \quad \sqrt{\epsilon}\Delta u_\epsilon^0 \longrightarrow 0 \text{ in } L^2(\Omega),$$

and with

(2.6) $$u_\epsilon = \frac{\partial u_\epsilon}{\partial v} = 0 \text{ on } \Sigma = \Gamma \times]0,T[.$$

223

Then, as $\epsilon \rightarrow 0$,

(2.7)
$$u_\epsilon \rightarrow u_0 \ in \ C([0,T];H_0^1(\Omega)) \ weak \ star,$$

$$u_\epsilon' \rightarrow u_0' \ in \ C([0,T];L^2(\Omega)) \ weak \ star$$

(2.8)
$$\{u_\epsilon(T),u_\epsilon'(T)\} \rightarrow \{u_0(T),u_0'(T)\} \ in \ H_0^1(\Omega) \times L^2(\Omega)$$

where

(2.9)
$$\begin{cases} u_0'' - \Delta u_0 = f, \\ \\ u_0(0) = u^0, \quad u_0'(0) = u^1, \\ \\ u_0 = 0 \ on \ \Sigma. \end{cases}$$

Moreover

(2.10)
$$-\sqrt{\epsilon}\Delta u_\epsilon|_\Sigma \rightarrow \frac{\partial u_0}{\partial v} \ in \ L^2(\Sigma).$$

Remark 2.1: Under the conditions of Theorem 2.1, it is well known that the solution u_ϵ, which exists and which is unique, satisfies

(2.11)
$$\begin{cases} u_\epsilon \in C([0,T];H_0^2(\Omega)), \\ \\ u_\epsilon' \in C([0,T];L^2(\Omega)). \end{cases}$$

This does *not* imply that $\Delta u_\epsilon|_\Sigma \in L^2(\Sigma)$. This property will be proven. It will be also proven that $\frac{\partial u_0}{\partial v} \in L^2(\Sigma)$, so that (2.10) makes sense. □

Remark 2.2: If we assume that (2.6) takes place in the *weak* topology, then we have conclusions similar to those of Theorem 2.1, by replacing in (2.8) and (2.10) the strong by the weak topology. □

The proof of Theorem 2.1 will be divided in several steps.

2.2   A priori estimates (I)

Let us set

(2.12) $E_\epsilon(t) = \frac{1}{2}[|\nabla u_\epsilon(t)|^2 + \epsilon|\Delta u_\epsilon(t)|^2 + |u_\epsilon'(t)|^2]$

where we use the following notations:

$$|\varphi|^2 = \int_\Omega \varphi^2 dx, \quad |\nabla\varphi|^2 = \int_\Omega \sum_i \left[\frac{\partial\varphi}{\partial x_i}\right]^2 dx.$$

We shall also set

$$(\varphi,\psi) = \int_\Omega \varphi\psi \ dx,$$

(2.13) $E_{0\epsilon} = \frac{1}{2}[|\nabla u_\epsilon^0|^2 + \epsilon|\Delta u_\epsilon^0|^2 + |u_\epsilon^1|^2].$

After multiplying (2.2) by u_ϵ', we obtain

(2.14) $E_\epsilon(t) = E_{0\epsilon} + \int_0^t (f_\epsilon, u_\epsilon') \ d\sigma.$

It immediately follows that, as $\epsilon \longrightarrow 0$,

$\{u_\epsilon, \sqrt{\epsilon}\Delta u_\epsilon, u_\epsilon'\}$ remains in a bounded set of

(2.15)

$L^\infty(0,T;H_0^1(\Omega)) \times L^\infty(0,T;L^2(\Omega)) \times L^\infty(0,T;L^2(\Omega)).$

We can extract a subsequence, still denoted by u_ϵ, such that (2.7) holds true, and such that u_0 is the solution of (2.9). Uniqueness of u_0 makes it unnecessary to extract a subsequence. Moreover

(2.16) $\sqrt{\epsilon}\Delta u_\epsilon \longrightarrow 0$ in $L^\infty(0,T;L^2(\Omega))$ weak star.

225

(We use indifferently L^∞ or C; the results are true in C).

According to (2.4) (2.5) we have

$$(2.17) \qquad E_{0\epsilon} \longrightarrow E_0 = \tfrac{1}{2}[\mid \nabla u^0 \mid^2 + \mid u^1 \mid^2].$$

Therefore (2.16) gives

$$(2.18) \qquad E_\epsilon(t) \longrightarrow E_0 + \int_0^t (f, u_0') \, d\sigma.$$

If we set

$$(2.19) \qquad E(t) = \tfrac{1}{2}[\mid \nabla u_0(t) \mid^2 + \mid u_0'(t) \mid^2]$$

then

$$E_0 + \int_0^t (f, u_0') \, d\sigma = E(t)$$

so that

$$(2.20) \qquad E_\epsilon(t) \longrightarrow E(t) \qquad \forall t,$$

and (2.8) follows, together with (2.16) in the *strong* topology.

2.3 A priori estimates (II)

We introduce functions h_k such that

$$(2.21) \qquad h_k \in C^2(\bar\Omega), \quad h_k = \nu_k \text{ on } \Gamma.$$

We multiply (2.2) by $h_k \dfrac{\partial u_\epsilon}{\partial x_k}$ (we use the summation convention of repeated indices) and we integrate by parts. We introduce the following notations:

226

$$\iint \cdots \int_{\Omega \times (0,T)} \phi \, dx \, dt = \iint \phi,$$

$$\int_{\Sigma} \phi \, d\Sigma = \int \phi$$

$$X_\epsilon = (u'_\epsilon(t), h_k \frac{\partial u_\epsilon}{\partial x_k}(t)) \Big|_0^T.$$

We have in this way:

$$X_\epsilon - \iint u'_\epsilon h_k \frac{\partial u'_\epsilon}{\partial x_k} + \epsilon \int \frac{\partial \Delta u_\epsilon}{\partial \nu} h_k \frac{\partial u_\epsilon}{\partial x_k} - \Delta u_\epsilon \frac{\partial}{\partial \nu}(h_k \frac{\partial u_\epsilon}{\partial x_k})$$

(2.22)
$$+ \epsilon \iint \Delta u_\epsilon \, \Delta(h_k \frac{\partial u_\epsilon}{\partial x_k}) + \iint \frac{\partial u_\epsilon}{\partial x_j} \frac{\partial}{\partial x_j}(h_k \frac{\partial u_\epsilon}{\partial x_k}) =$$

$$= \iint f_\epsilon \, h_k \frac{\partial u_\epsilon}{\partial x_k} .$$

Since $u_\epsilon = \dfrac{\partial u_\epsilon}{\partial \nu} = 0$ on Σ, the first surface integral in (2.22) equals

0 and the second one equals $-\epsilon \int (\Delta u_\epsilon)^2 = -\epsilon \int (\dfrac{\partial^2 u_\epsilon}{\partial \nu^2})^2.$ Then
(2.22) can be written

$$X_\epsilon - \iint \frac{h_k}{2} \frac{\partial}{\partial x_k} (u'_\epsilon)^2 - \epsilon \int (\Delta u_\epsilon)^2 +$$

$$+ \epsilon \iint \frac{h_k}{2} \frac{\partial}{\partial x_k} (\Delta u_\epsilon)^2 + 2\epsilon \iint \Delta u_\epsilon \frac{\partial h_k}{\partial x_j} \frac{\partial^2 u_\epsilon}{\partial x_j \partial x_k} +$$

(2.23)
$$+ \epsilon \iint \Delta h_k \, (\Delta u_\epsilon) \frac{\partial u_\epsilon}{\partial x_k} + \iint \frac{h_k}{2} \frac{\partial}{\partial x_k} |\nabla u_\epsilon|^2 +$$

$$+ \iint \frac{\partial h_k}{\partial x_j} \frac{\partial u_\epsilon}{\partial x_k} \frac{\partial u_\epsilon}{\partial x_j} = \iint f_\epsilon \, h_k \frac{\partial u_\epsilon}{\partial x_k}.$$

After integration by parts, it follows from (2.23) that

227

$$\frac{\epsilon}{2} \int (\Delta u_\epsilon)^2 = X_\epsilon + \iint \frac{1}{2} \text{ div } h\,[u_\epsilon'^2 - \epsilon(\Delta u_\epsilon)^2 - |\nabla u_\epsilon|^2] +$$

$$(2.24) \qquad + \iint \frac{\partial h_k}{\partial x_j}\,[2\epsilon\,\Delta u_\epsilon \frac{\partial^2 u_\epsilon}{\partial x_j \partial x_k} + \frac{\partial u_\epsilon}{\partial x_j}\frac{\partial u_\epsilon}{\partial x_k}] +$$

$$+ \iint \epsilon\,\Delta h_k\,\Delta u_\epsilon \frac{\partial u_\epsilon}{\partial x_k} - \iint f_\epsilon\,h_k \frac{\partial u_\epsilon}{\partial x_k}\,.$$

This identity implies that $\Delta u_\epsilon|_\Sigma \in L^2(\Sigma)$ (cf. Remark 2.1), and moreover that, as $\epsilon \longrightarrow 0$,

$$(2.25) \qquad \sqrt{\epsilon}\,\Delta u_\epsilon|_\Sigma \text{ remains in a bounded set of } L^2(\Sigma).$$

It follows from the results of Section 2.2 above that the right hand side of (2.24) converges toward

$$J = X + \iint \frac{1}{2}\text{ div } h\,[u_0'^2 - |\nabla u_0|^2] +$$

$$(2.26) \qquad + \iint \frac{\partial h_k}{\partial x_j}\frac{\partial u_0}{\partial x_j}\frac{\partial u_0}{\partial x_k} - \iint f\,h_k \frac{\partial u_0}{\partial x_k}\,.$$

But multiplying (2.9) by $h_k \dfrac{\partial u_0}{\partial x_k}$ and integrating by parts leads to

$$(2.27) \qquad \frac{1}{2} \int \left[\frac{\partial u_0}{\partial \nu}\right]^2 = J.$$

Remark 2.3: Identity (2.27) is a variant of the Rellich identity for elliptic equations. It has been used, for different purposes, in J.L. Lions [4]. It proves that $\dfrac{\partial u_0}{\partial \nu} \in L^2(\Sigma)$ (cf. Remark 2.1). □

Therefore we have proven that, as $\epsilon \longrightarrow 0$.

$$(2.28) \qquad \frac{\epsilon}{2} \int (\Delta u_\epsilon)^2 \longrightarrow \frac{1}{2} \int \left[\frac{\partial u_0}{\partial \nu}\right]^2.$$

Remark 2.4: In all of the results given in this section, T is given *arbitrarily*. □

2.4   Proof of (2.10) (I)

It follows from (2.25) that we can extract a subsequence, still denoted by u_ϵ, such that

(2.29) $-\sqrt{\epsilon} \ \Delta u_\epsilon |_\Sigma \longrightarrow x$ in $L^2(\Sigma)$ weakly.

The proof of Theorem 2.1 will be completed if we prove that

(2.30) $x = \dfrac{\partial u_0}{\partial \nu}$

since then (2.29) and (2.28) show strong convergence, i.e. (2.10).

Let us introduce now the Laplace transforms w_ϵ, w_0 of u_ϵ and u_0. We assume that f_ϵ and f are extended by 0 for $t > T$; then u_ϵ, u_0 are defined $\forall t > 0$, and we can introduce, $\forall p > 0$

(2.31) $w_\epsilon = \displaystyle\int_0^\infty e^{-pt} \ u_\epsilon(t)dt, \quad w_0 = \int_0^\infty e^{-pt} \ u_0(t)dt.$

The functions w_ϵ, w_0 are characterized by

$$\epsilon \Delta^2 w_\epsilon - \Delta w_\epsilon + p^2 w_\epsilon = F_\epsilon + u_\epsilon^1 + p u_\epsilon^0 \text{ in } \Omega,$$

(2.32)
$$w_\epsilon = \frac{\partial w_\epsilon}{\partial \nu} = 0 \text{ on } \Gamma$$

where

(2.33) $F_\epsilon = \displaystyle\int_0^\infty e^{-pt} \ f_\epsilon(t)dt,$

and

$$-\Delta w_0 + p^2 w_0 = F + u^1 + p u^0 \text{ in } \Omega,$$

(2.34)
$$w_0 = 0 \text{ on } \Gamma$$

where

$$F = \int_0^\infty e^{-pt} f(t)dt.$$

We have (2.30) if we can prove that

$$\int_0^\infty e^{-pt} x \, dt = \int_0^\infty e^{-pt} \frac{\partial u_0}{\partial \nu} \, dt \qquad \forall \, p > 0$$

and therefore it only remains to show that $\forall p > 0$ fixed,

(2.35) $\qquad\qquad -\sqrt{\epsilon} \; \Delta w_\epsilon |_\Gamma \longrightarrow \dfrac{\partial w_0}{\partial \nu}$ in $L^2(\Gamma)$ weakly.

2.5 Proof of (2.10) (II)

We now prove (2.35). We multiply (2.32) by $h_k \dfrac{\partial w_\epsilon}{\partial x_k}$, where the h_k's are chosen as in Section 2.3. It is now exactly Rellich's method! We obtain

$$\frac{\epsilon}{2} \int_\Gamma (\Delta w_\epsilon)^2 = \int_\Omega \frac{\text{div } h}{2} \, [-\epsilon(\Delta w_\epsilon)^2 - |\nabla w_\epsilon|^2 - p^2 w_\epsilon^2] +$$

(2.36) $$+ \int_\Omega \frac{\partial h_k}{\partial x_j} \, [2\epsilon \; \Delta w_\epsilon \frac{\partial^2 w_\epsilon}{\partial x_j \partial x_k} + \frac{\partial w_\epsilon}{\partial x_j} \frac{\partial w_\epsilon}{\partial x_k}] +$$

$$+ \epsilon \int_\Omega \Delta h_k \; \Delta w_\epsilon \frac{\partial w_\epsilon}{\partial x_k} - \int_\Omega (F + u_\epsilon^1 + p u_\epsilon^0) w_\epsilon .$$

It follows from (2.36) and from obvious estimates that, as $\epsilon \longrightarrow 0$,

$$\frac{\epsilon}{2} \int_\Gamma (\Delta w_\epsilon)^2 \longrightarrow \int_\Omega \frac{1}{2} \text{ div } h\left[-\mid \nabla w_0 \mid^2 - p^2 w_0^2\right] +$$

(2.37)

$$+ \int_\Omega \frac{\partial h_k}{\partial x_j} \frac{\partial w_0}{\partial x_j} \frac{\partial w_0}{\partial x_k} - \int_\Omega (F + u^1 + pu^0) w_0.$$

But the right hand side of (2.37) equals

$$\frac{1}{2} \int_\Gamma \left[\frac{\partial w_0}{\partial \nu}\right]^2$$

so that if (2.35) is proven, then in fact *strong* convergence will follow.

For the time being it follows from (2.37) that

(2.38) $\sqrt{\epsilon} \ \Delta w_\epsilon \mid_\Gamma$ is bounded in $L^2(\Gamma)$

so that we can assume that

(2.39) $-\sqrt{\epsilon} \ \Delta w_\epsilon \mid_\Gamma \longrightarrow \rho$ in $L^2(\Gamma)$ weakly

and it remains to show that

(2.40) $$\rho = \frac{\partial w_0}{\partial \nu} . \qquad \square$$

Let σ_0 be given arbitrarily, $\sigma_0 \in H^1(\Omega)$ and let σ_ϵ be defined by

(2.41)
$$\left| \begin{array}{l} -\epsilon \ \Delta\sigma_\epsilon + \sigma_\epsilon = \sigma_0 \text{ in } \Omega, \\[2mm] \sigma_\epsilon = 0 \text{ on } \Gamma. \end{array} \right.$$

We have $\sigma_\epsilon \in H^2(\Omega)$. We multiply (2.32) by σ_ϵ and we integrate by parts. We obtain

231

$$-\epsilon \int_{\Gamma} \Delta w_\epsilon \frac{\partial \mathcal{O}_\epsilon}{\partial \nu} + (\Delta w_\epsilon, \epsilon \Delta \mathcal{O}_\epsilon - \mathcal{O}_\epsilon) + p^2(w_\epsilon, \mathcal{O}_\epsilon) =$$

$$= (F + u_\epsilon^1 + pu_\epsilon^0, \mathcal{O}_\epsilon)$$

i.e.

(2.42)

$$\left| \int_{\Gamma} (-\sqrt{\epsilon}\ \Delta w_\epsilon)(\sqrt{\epsilon}\ \frac{\partial \mathcal{O}_\epsilon}{\partial \nu}) = (F + u_\epsilon^1 + pu_\epsilon^0, \mathcal{O}_\epsilon) - \right.$$

$$- (\nabla w_\epsilon, \nabla \mathcal{O}_0) - p^2(w_\epsilon, \mathcal{O}_\epsilon).$$

But we know that (in particular) $\mathcal{O}_\epsilon \longrightarrow \mathcal{O}_0$ in $L^2(\Omega)$ so that the right hand side of (2.42) converges to

$$(F + u^1 + pu^0, \mathcal{O}_0) - (\nabla w_0, \nabla \mathcal{O}_0) - p^2(w_0, \mathcal{O}_0)$$

and this quantity equals $-\int_{\Gamma} \frac{\partial w_0}{\partial \nu} \mathcal{O}_0$.

Let us admit for a moment that

(2.43)
$$\sqrt{\epsilon}\ \frac{\partial \mathcal{O}_\epsilon}{\partial \nu} \longrightarrow -\mathcal{O}_0 \text{ in } L^2(\Gamma).$$

Then

$$-\int_{\Gamma} p\mathcal{O}_0 = -\int_{\Gamma} \frac{\partial w_0}{\partial \nu} \mathcal{O}_0 \qquad \forall \mathcal{O}_0 \in H^1(\Omega)$$

hence (2.40) follows.

It only remains to verify (2.43), which is classical boundary layer theory. Indeed, if $\delta(x) = \text{distance}(x, \Gamma)$, we know that

$$\mathcal{O}_\epsilon(x) \simeq \mathcal{O}_0(1 - e^{-\delta(x)/\sqrt{\epsilon}})$$

in a neighborhood of Γ, hence (2.43) follows (we can assume here that $\mathcal{O}_0$ is as smooth as we please, so that in (2.43) the convergence could also take place in $C(\Gamma)$). □

3. **Singular perturbations. Non‑homogeneous problems**

   3.1 Setting of the problem

   We consider now the *non‑homogeneous problem*:

(3.1) $$w_\epsilon'' + \epsilon \, \Delta^2 w_\epsilon - \Delta w_\epsilon = 0 \text{ in } \Omega \times (0,T),$$

(3.2) $$w_\epsilon(0) = w_\epsilon^0, \ w_\epsilon'(0) = w_\epsilon^1 \text{ in } \Omega,$$

$$w_\epsilon = 0 \text{ on } \Sigma,$$

(3.3)

$$\frac{\partial w_\epsilon}{\partial \nu} = g_\epsilon \text{ on } \Sigma.$$

We assume that

(3.4)
$$w_\epsilon^0 \in L^2(\Omega), \ w_\epsilon^0 \longrightarrow w^0 \text{ in } L^2(\Omega) \text{ weakly},$$

$$w_\epsilon^1 \in H^{-1}(\Omega), \ w_\epsilon^1 \longrightarrow w^1 \text{ in } H^{-1}(\Omega) \text{ weakly}$$

and that

(3.5) $$-\sqrt{\epsilon} g_\epsilon \longrightarrow h \text{ in } L^2(\Sigma) \text{ weakly}.$$

We are going to prove the

Theorem 3.1: *Under hypotheses (3.4) (3.5) one has*

(3.6) $$w_\epsilon \longrightarrow w_0 \ in \ C([0,T];L^2(\Omega)) \ weak \ star$$

where w_0 is the solution of

233

$$
(3.7) \quad \left|
\begin{array}{l}
w_0'' - \Delta w_0 = 0, \\[2mm]
w_0(0) = w^0, \ w_0'(0) = w^1, \\[2mm]
w = h \ on \ \Sigma.
\end{array}
\right.
$$

Remark 3.1: As in Section 2--and contrary to what we shall assume in the following sections--T is given arbitrarily. □

Remark 3.2: Problem (3.1) (3.2) (3.3)--and problem (3.7) as well--is defined in the following weak sense:

Let f be given in $L^1(0,T;L^2(\Omega))$. Let $\emptyset_\epsilon$ be given by

$$
(3.8) \quad \left|
\begin{array}{c}
\emptyset_\epsilon'' + \epsilon \Delta^2 \emptyset_\epsilon - \Delta \emptyset_\epsilon = f \ in \ \Omega \times \,]0,T[, \\[3mm]
\emptyset_\epsilon(T) = \emptyset_\epsilon'(T) = 0, \\[3mm]
\emptyset_\epsilon = \dfrac{\partial \emptyset_\epsilon}{\partial \nu} = 0 \ on \ \Sigma.
\end{array}
\right.
$$

Because of time reversibility we can apply the results of Section 2.1.

If we multiply (3.1) *formally* by $\emptyset_\epsilon$, we obtain after integrating by parts and using (3.2) (3.3):

$$
(3.9) \quad \int_{\Omega \times (0,T)} w_\epsilon f = (w_\epsilon^1, \emptyset_\epsilon(0)) - (w_\epsilon^0, \emptyset_\epsilon'(0)) - \int_\Sigma \sqrt{\epsilon} g_\epsilon \ \sqrt{\epsilon} \ \Delta \emptyset_\epsilon.
$$

As it is usual in problems of this type (cf. Lions-Magenes [1]) we *define* w_ϵ by (3.9); we obtain in this way $w_\epsilon \in L^\infty(0,T;L^2(\Omega))$ and we verify next that in fact $w_\epsilon \in C([0,T];L^2(\Omega))$. □

234

3.2   Proof of Theorem 3.1

According to Section 2, we know that

(3.10) $$\sigma_\epsilon \longrightarrow \sigma_0 \text{ in } C([0,T];H_0^1(\Omega)) \text{ weak star,}$$

(3.11) $$\{\sigma_\epsilon(0),\sigma_\epsilon'(0)\} \longrightarrow \{\sigma_0(0),\sigma_0'(0)\} \text{ in } H_0^1(\Omega) \times L^2(\Omega),$$

(3.12) $$-\sqrt{\epsilon}\ \Delta\sigma_\epsilon|_\Sigma \longrightarrow \frac{\partial\sigma_0}{\partial\nu} \text{ in } L^2(\Sigma_i)$$

where

(3.13) $$\left|\begin{array}{l} \sigma_0'' - \Delta\sigma_0 = f, \\[2mm] \sigma_0(T) = \sigma_0'(T) = 0, \\[2mm] \dfrac{\partial\sigma_0}{\partial\nu} = 0 \text{ on } \Sigma. \end{array}\right.$$

The right hand side of (3.9) converges to

(3.14) $$(w^1,\sigma_0(0)) - (w^0,\sigma_0'(0)) - \int_\Sigma h\,\frac{\partial\sigma_0}{\partial\nu}.$$

We have that w_ϵ remains in a bounded subset of $L^\infty(0,T;L^2(\Omega))$. We can assume that one has (3.6), where

$$\int_{\Omega\times(0,T)} w_0 f = (w^1,\sigma_0(0)) - (w^0,\sigma_0'(0)) - \int_\Sigma h\,\frac{\partial\sigma_0}{\partial\nu}.$$

This is the definition of the weak solution of (3.7), hence the theorem follows.   □

4.   Hilbert structure on initial datas associated with a uniqueness result

4.1   Principle of the general method

We now make explicit the general method introduced in J.L. Lions [1] in the present situation. Cf. J.L. Lions [2], method HUM. We use the notation of the Introduction.

For φ^0, φ^1 given in Ω (without, for the time being, making precise the function spaces where the φ^i's belong to), we define φ_ϵ by

(4.1)
$$\begin{vmatrix} \varphi_\epsilon'' + \epsilon \, \Delta^2 \varphi_\epsilon - \Delta \varphi_\epsilon = 0 \text{ in } \Omega \times (0,T), \\ \\ \varphi_\epsilon(0) = \varphi^0, \; \varphi_\epsilon'(0) = \varphi^1, \\ \\ \varphi_\epsilon = \dfrac{\partial \varphi_\epsilon}{\partial \nu} = 0 \text{ on } \Sigma. \end{vmatrix}$$

We then define ψ_ϵ by

(4.2)
$$\begin{vmatrix} \psi_\epsilon'' + \epsilon \, \Delta^2 \psi_\epsilon - \Delta \psi_\epsilon = 0 \text{ in } \Omega \times (0,T), \\ \\ \psi_\epsilon(T) = \psi_\epsilon'(T) = 0, \\ \\ \psi_\epsilon = 0 \text{ on } \Sigma, \\ \\ \dfrac{\partial \psi_\epsilon}{\partial \nu} = \begin{vmatrix} \Delta \varphi_\epsilon & \text{on } \Sigma(x^0) \\ 0 & \text{on } \Sigma^*(x^0) \end{vmatrix}. \end{vmatrix}$$

We then *define* $\Lambda_\epsilon \{\varphi^0, \varphi^1\}$ by

(4.3)
$$\Lambda_\epsilon \{\varphi^0, \varphi^1\} = \{\psi_\epsilon'(0), -\psi_\epsilon(0)\}.$$

We have in this way defined a mapping; we have now to make precise the function spaces associated with Λ_ϵ. □

In a formal way (for the time being), the general idea is to show that

(i)   provided T is large enough,

(ii)   provided the function spaces are conveniently defined,

Λ_ϵ is *invertible.*

Then we "solve"

(4.4)
$$\Lambda_\epsilon \{\varphi^0_\epsilon, \varphi^1_\epsilon\} = \{y^1, -y^0\}.$$

This defines $\{\varphi^0_\epsilon, \varphi^1_\epsilon\}$. Equations (4.1) then define φ_ϵ. If we *define*

(4.5)
$$v_\epsilon = \Delta\varphi_\epsilon \text{ on } \Sigma(x^0)$$

then v_ϵ solves *the exact controllability problem*. If y_ϵ denotes the solution of (1.6) (1.10) (1.11), then $y_\epsilon(T) = y'_\epsilon(T) = 0$.

We now make all this precise--and we shall let $\epsilon \longrightarrow 0$ (and prove the main theorem) in the following Section 5.

4.2   The function space F

Let us proceed formally, all this being made rigorous later on. We multiply (4.2) by φ_ϵ and we integrate by parts. We find

(4.6)
$$<\Lambda_\epsilon\{\varphi^0,\varphi^1\},\{\varphi^0,\varphi^1\}> = \int_{\Sigma(x^0)} \epsilon(\Delta\varphi_\epsilon)^2.$$

If we can show that

(4.7)
$$\left(\int_{\Sigma(x^0)} \epsilon(\Delta\varphi_\epsilon)^2 \right)^{1/2}$$

defines a norm on the space of initial datas $\{\varphi^0,\varphi^1\}$, then *we define* F as the completion of, say, $\mathcal{D}(\Omega) \times \mathcal{D}(\Omega)$ ($\mathcal{D}(\Omega) = C^\infty$ functions with compact support in Ω).

One verifies then that

(4.8)
$$\{\psi'_\epsilon(0), -\psi_\epsilon(0)\} \in F' = \text{dual space of } F$$

(we identify $L^2(\Omega)$ with its dual, so that--in general--F' cannot be identified with its dual).

It then follows from (4.6) and from *the definition of* F

237

that

(4.9) Λ_ϵ *defines an isomorphism from* F *onto* F'.

Then the procedure of Section 4.1 becomes correct, provided

(4.10) $\{y^1, y^0\} \in$ F'.

Remark 4.1: Saying that (4.7) is a norm amounts *to a uniqueness theorem*: if φ_ϵ satisfies

(4.11)
$$\begin{vmatrix} \varphi_\epsilon \in C([0,T];H_0^2(\Omega)), \\[2mm] \varphi_\epsilon' \in C([0,T];L^2(\Omega)), \\[2mm] \varphi_\epsilon'' + \epsilon \, \Delta^2 \varphi_\epsilon - \Delta \varphi_\epsilon = 0 \text{ in } \Omega \times (0,T), \end{vmatrix}$$

(4.12) $\Delta \varphi_\epsilon = 0$ on $\Sigma(x^0)$,

(4.13)   T is large enough

then $\varphi_\epsilon = 0$.

Such a result will be proven below.   □

Remark 4.2:   It is quite clear that the method indicated above is general.   "Something" will work--as far as exact controllability is concerned--each time we shall have a uniqueness theorem.   It is a classical remark (cf. D.L. Russell [1]) that uniqueness implies *weak* controllability.   What we use here is the fact that, provided the function spaces are suitably chosen, uniqueness implies *strong* controllability.   □

Remark 4.3:   Of course the next step is to obtain information on F (a priori this space *could* depend on ϵ).   We shall prove:

Theorem 4.1:   *There is a* $T_0 > 0$ *independent of* ϵ, *such*

that for $T > T_0$, *(4.7) is a norm which is equivalent to*
$(\|\varphi^0\|^2_{H^2_0(\Omega)} + \|\varphi^1\|^2_{L^2(\Omega)})^{1/2}$.

In other words

(4.14)
$$F = H^2_0(\Omega) \times L^2(\Omega).$$

Corollary:

(4.15)
$$F' = H^{-2}(\Omega) \times L^2(\Omega)$$

so that, for every $\epsilon > 0$, *the system (1.6) (1.10) (1.11) is exactly controllable, starting with* $y^0 \in L^2(\Omega)$, $y^1 \in H^{-1}(\Omega)$. □

We can give estimates on T_0 appearing in Theorem 4.1.
Let λ^2_0 be the *first* eigenvalue of

(4.16)
$$-\Delta w = \lambda^2_0 w, \; w = 0 \text{ on } \Gamma.$$

Then

(4.17)
$$|w| \leqslant \frac{1}{\lambda_0} |\nabla w| \qquad \forall w \in H^1_0(\Omega).$$

Let us define

(4.18)
$$R(x^0) = \text{distance from } x^0 \text{ to } \Gamma$$

and let us set

(4.19)
$$T(x^0) = 2R(x^0) + \frac{n-1}{\lambda_0} .$$

Then

(4.20)
$$T_0 \leqslant T(x^0).$$

239

Added in Proof

One can take $T(x^0) = 2R(x^0)$--cf. V. Komornik, to appear.

4.3 Proof of Theorem 4.1

Let us define

$$(4.21) \quad \begin{array}{l} E_{0\epsilon} = \frac{1}{2}[\,|\,\nabla\varphi^0\,|^2 + \epsilon\,|\,\Delta\varphi^0\,|^2 + |\,\varphi^1\,|^2], \\[2mm] E_0 = \frac{1}{2}[\,|\,\nabla\varphi^0\,|^2 + |\,\varphi^1\,|^2]. \end{array}$$

We use (2.24) with $u_\epsilon = \varphi_\epsilon$, $f_\epsilon = 0$. It follows that there exists a constant c, *independent of* T, such that

$$(4.22) \quad \epsilon \int_{\Gamma\times(0,T)} (\Delta\varphi_\epsilon)^2 \leqslant c\,T\,E_{0\epsilon}.$$

This is valid for $T > 0$ arbitrarily small.

We now prove a "reverse" inequality, only valid for T large enough. $\square$

With definition (4.9) for $m_k(x)$, we multiply equation (4.11) by $m_k \dfrac{\partial\varphi_\epsilon}{\partial x_k}$. Let us drop the index "ϵ" in the following computation. We use notations similar to those of Section 2. We set

$$(4.23) \quad X = (\varphi'(t), m_k \frac{\partial\varphi(t)}{\partial x_k})\Big|_0^T.$$

We obtain

$$X - \iint \frac{m_k}{2}\frac{\partial}{\partial x_k}(\varphi'^2) - \epsilon\int \Delta\varphi\frac{\partial}{\partial\nu}(m_k\frac{\partial\varphi}{\partial x_k}) +$$

$$+ \epsilon\iint \Delta\varphi\,\Delta(m_k\frac{\partial\varphi}{\partial x_k}) + \iint \frac{\partial\varphi}{\partial x_j}\frac{\partial}{\partial x_j}(m_k\frac{\partial\varphi}{\partial x_k}) = 0$$

hence

$$(4.24) \quad X + \frac{n}{2} \iint \varphi'^2 - \epsilon \int (m_k v_k)(\Delta\varphi)^2 + \epsilon \iint \frac{m_k}{2} \frac{\partial}{\partial x_k}(\Delta\varphi)^2 +$$

$$+ 2\epsilon \iint (\Delta\varphi)^2 + \iint \frac{m_k}{2} \frac{\partial}{\partial x_k} |\nabla\varphi|^2 + \iint |\nabla\varphi|^2 = 0.$$

It follows from (4.24) that

$$X + \frac{n-1}{2} \iint \varphi'^2 - \epsilon(\Delta\varphi)^2 - |\nabla\varphi|^2 +$$

$$(4.25) \quad + \frac{1}{2} \iint \varphi'^2 + \epsilon(\Delta\varphi)^2 + |\nabla\varphi|^2 + \epsilon \iint (\Delta\varphi)^2$$

$$- \frac{\epsilon}{2} \int (m_k v_k)(\Delta\varphi)^2 = 0.$$

But

$$(4.26) \quad \iint \varphi'^2 - \epsilon(\Delta\varphi)^2 - |\nabla\varphi|^2 = (\varphi'(t), \varphi(t)) \Big|_0^T = Y.$$

We rewrite (4.25) as follows:

$$T E_{0\epsilon} + \epsilon \iint (\Delta\varphi)^2 - \frac{\epsilon}{2} \int (m_k v_k)(\Delta\varphi)^2 =$$

$$(4.27) \quad = -X - \frac{n-1}{2} Y + \frac{\epsilon}{2} \int_{\Sigma(x^0)} m_k v_k (\Delta\varphi)^2$$

(we recall that $\Sigma(x^0)$ (resp. $\Sigma^*(x^0)$) is the part of Σ where $m_k v_k \geq 0$ (resp. ≤ 0)).

It follows from (4.27) that

$$(4.28) \quad T E_{0\epsilon} \leq |X| + \frac{n-1}{2} |Y| + \epsilon \frac{R(x^0)}{2} \int_{\Sigma(x^0)} (\Delta\varphi)^2.$$

We observe now that

$$|(\varphi'(t), m_k \frac{\partial\varphi(t)}{\partial x_k})| \leq R(x^0) |\varphi'(t)| \, |\nabla\varphi(t)| \leq R(x^0) E_{0\epsilon}$$

so that

(4.29)
$$|X| \leq 2R(x^0) \, E_{0\epsilon}.$$

Next

$$|(\varphi'(t),\varphi(t))| \leq |\varphi'(t)| \frac{1}{\lambda_0} |\nabla\varphi(t)| \leq \frac{1}{\lambda_0} E_{0\epsilon}$$

so that

(4.30)
$$|Y| \leq \frac{2}{\lambda_0} E_{0\epsilon}.$$

Therefore

(4.31)
$$T \, E_{0\epsilon} \leq (2R(x^0) + \frac{n-1}{\lambda_0})E_{0\epsilon} + \frac{\epsilon R(x^0)}{2} \int_{\Sigma(x^0)} (\Delta\varphi)^2.$$

Conclusion (we use again the index ϵ)

(4.32)
$$\epsilon \int_{\Sigma(x^0)} (\Delta\varphi_\epsilon)^2 \geq \frac{2}{R(x^0)} (T - T(x^0)) \, E_{0\epsilon}, \quad T > T(x^0).$$

Theorem 4.1 follows. □

Remark 4.4: Suppose that Ω is convex (but the remark is quite general!). Then if $x^0 \in \Omega$, $\Gamma(x^0) = \Gamma$ $\quad \forall x^0$, so that one can then use *all* boundary Γ. The "best" formula for $T(x^0)$ is then obtained by taking x^0 at the "center" of Ω, so that

$$2R(x^0) = \text{diameter } (\Omega).$$

Then we can exact control the system

$$y_\epsilon'' + \epsilon\Delta^2 y_\epsilon - \Delta y_\epsilon = 0,$$

starting from $L^2(\Omega) \times H^{-2}(\Omega)$, with

$$y_\epsilon = 0, \quad \frac{\partial y_\epsilon}{\partial \nu} = v_\epsilon \quad on \; the \; whole \; \Sigma$$

242

in a time $T > T_0$, $T_0 \leqslant$ diameter (Ω).

If $x^0 \in C\ \bar{\Omega}$, then $\Gamma(x^0)$ is *part* of Γ. This part "decreases" as x^0 goes farther from Ω, but then $T(x^0)$ *increases*: one has to act *longer* if acting only on *part* of Γ.   □

Remark 4.5:   An estimate of the type (4.32) has been given for hyperbolic equations by Lop Fat Ho [1].   □

Remark 4.6:   Among all controls $v \in L^2(\Sigma(x^0))$ which drive y_ϵ to rest at time T (there are an infinite number of such v's when $T > T_0$), the control v_ϵ constructed above is the one which *minimizes* $\int_{\Sigma(x^0)} v^2$.   □

Remark 4.7:   Let T_0^* be such that uniqueness as in Remark 4.1 holds true. *Then we can take* $T_0 = T_0^*$ *in Theorem* 4.1. To prove that, one estimates differently $|Y|$. One has, for every $\eta > 0$,

$$|Y| \leqslant \eta\ E_{0\epsilon} + c(\eta)\ \|\varphi_\epsilon\|^2_{L^\infty(0,T;L^2(\Omega))}$$

and one proves that there is a constant c such that

$$\|\varphi_\epsilon\|^2_{L^\infty(0,T;L^2(\Omega))} \leqslant c\ \epsilon \int_{\Sigma(x^0)} (\Delta\varphi_\epsilon)^2$$

by a compactness argument ([1]).   □

5.   **Proof of the main result**

We use Λ_ϵ as defined by (4.4).

([1])This remark is due to Pierre Louis Lions.

We already know that Λ_ϵ is an isomorphism from $F = H_0^2(\Omega) \times L^2(\Omega)$ into its dual, $\forall \epsilon > 0$, $\forall T > T_0$. This shows Theorem 1.1.

Let us now assume that

(5.1)
$$y^0 \in L^2(\Omega), \; y^1 \in H^{-1}(\Omega) \; (\text{and } not \; H^{-2}(\Omega)).$$

We solve

(5.2)
$$\Lambda_\epsilon \{\varphi_\epsilon^0, \varphi_\epsilon^1\} = \{y^1, -y^0\}.$$

Taking the scalar product with $\{\varphi_\epsilon^0, \varphi_\epsilon^1\}$, we obtain

(5.3)
$$\| \{\varphi_\epsilon^0, \varphi_\epsilon^1\} \|_F^2 = (y^1, \varphi_\epsilon^0) - (y^0, \varphi_\epsilon^1) \leqslant$$
$$\leqslant \|y^1\|_{H^{-1}(\Omega)} \; |\nabla \varphi_\epsilon^0| + |y^0| \; |\varphi_\epsilon^1|.$$

Assuming $T > T(x^0)$ and using Theorem 4.1, (and more precisely (4.31)) we have:

(5.4)
$$|\nabla \varphi_\epsilon^0|^2 + \epsilon |\Delta \varphi_\epsilon^0|^2 + |\varphi_\epsilon^1|^2 \leqslant \text{constant}.$$

Therefore we can extract a subsequence such that

(5.5)
$$\varphi_\epsilon^0 \longrightarrow \varphi^0 \text{ in } H_0^1(\Omega) \text{ weakly,}$$
$$\varphi_\epsilon^1 \longrightarrow \varphi^1 \text{ in } L^2(\Omega) \text{ weakly,}$$
$$\sqrt{\epsilon} \Delta \varphi_\epsilon^0 \longrightarrow 0 \text{ in } L^2(\Omega) \text{ weakly.}$$

We use Theorem 2.1 in its "weak" form (cf. Remark 2.2). It follows that

$$\varphi_\epsilon \longrightarrow \varphi_0 \text{ in } C([0,T];H_0^1(\Omega)) \text{ weak star,}$$

(5.6)

$$\varphi_\epsilon' \longrightarrow \varphi_0' \text{ in } C([0,T];L^2(\Omega)) \text{ weak star,}$$

(5.7)
$$-\sqrt{\epsilon}\Delta\varphi_\epsilon|_\Sigma \longrightarrow \frac{\partial\varphi_0}{\partial\nu} \text{ in } L^2(\Sigma) \text{ weakly}$$

where

$$\varphi_0'' - \Delta\varphi_0 = 0,$$

(5.8)
$$\varphi_0(0) = \varphi^0, \quad \varphi_0'(0) = \varphi^1,$$

$$\varphi_0 = 0 \text{ on } \Sigma.$$

We consider now the equation (4.2) and we apply Theorem 3.1 with reversed time and with

(5.9)
$$g_\epsilon = \begin{vmatrix} \Delta\varphi_\epsilon & \text{on } \Sigma(x^0) \\ 0 & \text{on } \Sigma^*(x^0) . \end{vmatrix}$$

This is possible since according to (5.7)

(5.10)
$$-\sqrt{\epsilon}g_\epsilon \longrightarrow h = \begin{vmatrix} \dfrac{\partial\varphi_0}{\partial\nu} & \text{on } \Sigma(x^0) \\ 0 & \text{on } \Sigma^*(x^0) \end{vmatrix} \text{ in } L^2(\Sigma) \text{ weakly.}$$

It follows that

(5.11)
$$\psi_\epsilon \longrightarrow \psi_0 \text{ in } C([0,T];L^2(\Omega)) \text{ weak star}$$

where ψ_0 is given by

$$\psi_0'' - \Delta\psi_0 = 0,$$

(5.12)
$$\psi_0(T) = \psi_0'(T) = 0,$$

$$\psi_0 = \begin{vmatrix} \dfrac{\partial\psi_0}{\partial\nu} & \text{on } \Sigma(x^0) \\ 0 & \text{on } \Sigma^*(x^0). \end{vmatrix}$$

Therefore, if we define

(5.13)
$$v_0 = \frac{\partial v_0}{\partial\nu} \text{ on } \Sigma(x^0),$$

v_0 is the control which drives system (1.16) to rest at time T and which minimizes

(5.14)
$$\int_{\Sigma(x^0)} v^2.$$

The control v_ϵ being given by

$$v_\epsilon = \Delta\varphi_\epsilon \text{ on } \Sigma(x^0),$$

the proof of Theorem 1.2 is completed. □

Added in proof (December 4, 1986)

1)   Following a remark of V. KOMORNIK (to appear in the
C.R.A.S., 1987) one can take in (4.19) (and in the following formula)
$T(x^0) = 2R(x^0)$.

2)   One can also obtain the main result of this paper by using
duality theory.   The "dual proof" is given in J.L. LIONS,
Colloquium on Singular Perturbations, Paris, November 1986.

246

Bibliography

M. Avellaneda	[1]	To appear.
L.F. Ho	[1]	C.R.A.S. 302 (1986).
J.L. Lions	[1]	C.R.A.S. 302 (1986) p. 471-475.
	[2]	J. von Neumann Lecture. SIAM. 1986. To appear in SIAM Review.
	[3]	Lecture Notes of College de France. 1986/87.
	[4]	Contrôle des systèmes singuliers. Dunod. Paris. 1983.
J.L. Lions and E. Magenes	[1]	Problèmes aux limites non homogènes. Dunod. Paris. 1968. Vol. 2.
D.L. Russell	[1]	Studies in Applied Math. 52, 1973, p. 189-211.
	[2]	SIAM Review 20 (1978), p. 639-739.

TRANSONIC FLOW AND COMPENSATED COMPACTNESS

By

Cathleen S. Morawetz

1.   Introduction

The problem of finding steady flow past an airfoil is an old problem going back to the time of Lord Rayleigh. The understanding that there was a difficulty connected to the transition from subsonic flow to supersonic flow must surely, however, be attributed to Chaplygin [1], whose famous thesis describing solutions of the equations with such transitions was written in 1904.   The first mathematical study of such transitions which force a change of type for the differential equations from elliptic to hyperbolic began with the work of Tricomi [2] in 1923.   In 1930 at the International Mechanics Congress, Busemann [3] with wind tunnel data, and G.I. Taylor [4] with some computations, presented opposing views of the airfoil problem, the former suggesting that perhaps no steady flow existed and the latter than a series expansion in Mach number gave no evidence of a breakdown when the type changed.

What has emerged eventually is that no smooth irrotational two-dimensional flow can be expected, Morawetz [5], and that one must seek a weak solution. Following the work of Lax these solutions should satisfy an entropy condition.   This means that density must increase in the direction of flow or that in some sense the gradient of the density in the direction of flow is bounded from below.

The boundary value problem is that the potential ϕ satisfies weakly:

(i)   The conservation of mass: div $\rho_B \nabla\phi = 0$ in Ω

(ii)   Flow is tangential on the airfoil: $\partial\phi/\partial n = 0$ on $\partial\Omega$

Research supported in part by NSF Grant DMS-8120790.

(iii)   At infinity, the speed is prescribed: $\nabla\phi = (q_\infty, 0)$

(iv)   Entropy condition $\nabla\rho_B \cdot \nabla\phi > -\infty$.

Here $\rho_B = \rho_B(|\nabla\phi|)$ is the density given by Bernoulli's law which we write in the form

(v) $\rho_B \, q \, dq + dp = 0$, $\rho_B(q) = 0$ for $q \geq q_*$

where $p = p(\rho_B)$ and $q = |\nabla\phi|$.

In terms of components of velocity $(u,v) = \nabla\phi$ we find, with $dp/d\rho = c^2$,

$$(c^2-u^2)\phi_{xx} - 2uv\phi_{xy} + (c^2-v^2)\phi_{yy} = 0$$

which is elliptic for $u^2 + v^2 < c^2$, hyperbolic for $u^2 + v^2 > c^2$.

Finally, as in incompressible flow, a requirement that the flow is smooth at the trailing edge cusp leads to a well-posed elliptic problem if the speed q_∞ is sufficiently small, see Shiffman [6], Bers [7].

For flows involving shocks and transonic regions, numerical solutions involving artificial viscosity in a variety of ways, see Bauer, Garabedian, Korn [8] and Jameson [9], suggest that by adding viscous terms the b.v.p. described above would be soluble and by letting the viscosity tend to zero we might obtain an existence theorem.

2.   Viscous problem

We introduce a viscosity in the Euler equations of momentum but retain the notion that the flow is irrotational. Thus,

$$uu_x + vu_y = \nu \, \Delta u - p_x/\rho$$

$$uv_x + vv_y = -\nu \, \Delta v - p_y/\rho$$

which leads to

$$\nu \; \Delta\phi = |\nabla\phi|^2 - q_B^2(\rho)$$

where $q_B(\rho)$ is defined by $q_B(\rho_B(q)) = q$ for $\rho_B \neq 0$ and $q_B(\rho) = 0$ for $\rho \geqslant \rho_{\text{stagnation}}$.

Coupling the equation for $\Delta\phi$ with div $\rho \, \nabla\phi = 0$ we have the third order system, $\nu > 0$,

(2.1)

$$\nu \; \nabla\rho \; \cdot \; \nabla\phi = -\rho(|\nabla\phi|^2 - q_B^2(\rho))$$

$$\text{div} \; \rho \; \nabla\phi = 0$$

and we add the boundary condition

$$\rho \longrightarrow \rho_B(q_\infty) \; \text{for} \; \phi \longrightarrow -\infty.$$

To avoid dealing with an infinite domain we consider a symmetric airfoil in a finite domain. To cope with the problem of cavitation ($\rho = 0$) and infinite velocity we consider a free boundary value problem. Thus the problem we are attacking for $\nu > 0$ is to solve (2.1) in Ω lying in $0 \leqslant x \leqslant a$, $0 \leqslant Y(x) \leqslant y \leqslant b$. The boundary of Ω consists of $x = 0$, $0 \leqslant y \leqslant b$, $y = b$, $0 \leqslant x \leqslant a$; $x = a$, $b_* \leqslant y \leqslant b$ with $b_* \geqslant 0$; $y = Y(x)$, $0 \leqslant x \leqslant a_*$ with $a_* \leqslant a$ and Γ where Γ is a free boundary joining $(a_*, Y(a_*))$ to (a, b_*).

The boundary conditions for ϕ, ρ on $\partial\Omega$ are

on $x = 0$, $\phi = 0$, $\rho = \rho_0$,

on $y = b$, $\partial\phi/\partial n = 0$,

(2.2)   on $y = Y$, $\partial\phi/\partial n = 0$, $|\nabla\phi| \leqslant q_*$,

on $x = a$, $\phi = A$, $|\nabla\phi| \leqslant q_*$,

on Γ, $\phi = A$, $|\nabla\phi| = q_*$.

The value of $A/a < q_*$ replaces prescribing the speed at ∞; ρ_0 is some given value of density.

This model was introduced in Morawetz [10] and the results presented here are the same. The key point of the proof which separates the difficulties of the transition from subsonic to supersonic has been simplified.

It is not obvious that the solution of (2.1) and (2.2) can be established for $\nu > 0$ and $A/a < q_*$. An alternative possibility where existence might be established involves a second smoothing of ρ not only backwards along the streamlines but also symmetrically in the orthogonal direction. This is technically easy if we have no free boundary and is currently under investigation if we do. Perhaps there are many possible smoothings but of course they must be restricted to those that give the right entropy condition.

For the purpose of this paper we make

Assumption 1: There exists a solution for $\nu > 0$ of (2.1), (2.2) with ϕ, ρ continuous.

A number of estimates independent of ν can be derived easily using (2.1) along with the maximum principle for elliptic equations and the similarity principle of Bers and Nirenberg:

$$0 \leqslant \phi \leqslant A$$

$$\rho \neq 0, \qquad 0 < \rho < \rho_{stagnation}$$

(2.3)

$$|\nabla \phi| = |w| \leqslant q_*$$

$$0 \leqslant \psi \leqslant \rho_{stagnation} q_*^2 ab$$

where the stream function ψ is defined by

$$\rho \ d\phi + i \ d\psi = \rho w \ dz$$

(2.3a)

$$z = x + iy, \ w = u - iv.$$

A more difficult estimate shows that $q \longrightarrow q_B(\rho)$ as $\nu \longrightarrow 0$. Let $w = \exp(\sigma - i\theta)$, $s = \log q_B(\rho)$. Then

(2.4)
$$\int_{\Omega} e^{2\sigma} |\sigma - s|^2 \ dx \ dy \leqslant M\nu$$

where M is independent of ν.

From the uniform boundedness of $|\nabla\phi|$ there exists a subsequence such that as $\nu \longrightarrow 0$, ϕ tends to a limit which we continue to use ϕ to denote. By integrating by parts for χ of compact support,

$$\int \chi \ \text{div} \ \rho_B \ \nabla\phi \ | dx | = \int (\rho - \rho_B) \ \nabla\phi \cdot \nabla\chi \ dx \longrightarrow 0$$

by (2.4).

Thus div $\rho_B \nabla\phi$ tends weakly to zero as $\nu \longrightarrow 0$.

In fact all bounded functions of w, i.e. of $\nabla\phi$, converge weakly but what we have to show to prove the desired existence theorem is that

(2.5)
$$\text{w.}\ell. \ \rho_B(|\nabla\phi|) = \rho_B(\text{w.}\ell. \ |\nabla\phi|)$$

where w.ℓ. = weak limit.

To do this the only tool at hand is the method of compensated compactness introduced by Tartar and Murat, see [11]. For this we need an additional estimate which is not hard to derive once we have (2.5). This is

(2.6)
$$\int e^{2\sigma} (|\nabla\phi|^2 + |\nabla\theta|^2) \ |dx| \leqslant M\nu^{-1}.$$

The details to obtain the estimates are to be found in Morawetz [10].

3. Application of the method of compensated compactness

An entropy pair as defined in the ϕ,ψ-plane is a pair of functions $f(\theta,\sigma)$, $g(\theta,\sigma)$ bounded uniformly in L^2 with respect to ν and such that $f_\phi + g_\psi$ remains compact in $H_{loc}^{-1}(\phi,\psi)$ as $\nu \longrightarrow 0$.

Every entropy pair has a weak limit but, by the div-curl lemma of Tartar and Murat, if f_1,g_1 and f_2,g_2 are two entropy pairs then

$$\text{w.}\ell.(f_1g_2-g_1f_2) = (\text{w.}\ell.\ f_1)(\text{w.}\ell.\ g_2) - (\text{w.}\ell.\ g_1)(\text{w.}\ell.\ f_2).$$

We have a useful representation of w.ℓ. using the Young measure

$$\text{w.}\ell.\ f(\theta(x,y),\sigma(x,y)) = \int_{\bar\Omega \subset \{\theta,\sigma\}} f(\theta,\sigma)\ dm(\theta,\sigma;x,y)$$

where dm is a probability measure.

We may write the div-curl lemma then as

$$\iint_{\substack{\theta,\sigma\subset\Omega_- \\ \theta',\sigma'\subset\Omega'}} (f_1g_2' - f_2g_1')\ dm\ dm' = \iint (f_1g_2 - f_2g_1)\ dm\ dm'$$

Here prime means the argument is θ',σ'.

Our object is to show through the div-curl lemma that $dm = \delta\ d\theta\ d\sigma$ where δ is a Dirac delta-function. This idea was used by Tartar [11] for a sclar problem and by Di Perna [12] for purely hyperbolic problems. Following him we construct two families of entropy pairs depending on a discrete parameter n. We then let $n \longrightarrow \infty$ and conclude by some delicate arguments that the only possible measures are Dirac.

Then $\quad$ w.ℓ. $\rho_B(|\nabla\phi|) \quad = \rho_B(\text{w.}\ell.\ |\nabla\phi|) \quad$ and w.ℓ. div $\rho_B(|\nabla\phi|)$w.ℓ. $\nabla\phi = 0$ almost everywhere as required.

To find entropy pairs we note that $f_\phi + g_\psi = f_\sigma\sigma_\phi + f_\theta\theta_\phi + g_\sigma\sigma_\psi + g_\theta\theta_\psi$. In the right hand side we use the differential equations for σ and θ. These can be derived from (2.3a) and are $\rho \sigma_\psi - \theta_\phi = 0$ and $\tau \sigma_\phi + \theta_\psi + \tau_\phi = 0$. One then finds that if

$$\rho_B f_\theta + g_\sigma = 0, \quad f_\sigma + (\rho_B^{-1} + (\rho_B^{-1})_\sigma) g_\theta = 0,$$

i.e. satisfy the hodograph equations, then f and g are entropy pairs because of the estimates (2.4) and (2.6).

Setting $g = -H_\theta$, $f = \tau_B(\sigma) H_\sigma$, $d\mu = \rho \, d\sigma$, $\mu = 0$ at sonic speed we find the Tricomi-like equation,

$$\tau^2 (1 - M^2) H_{\theta\theta} + H_{\mu\mu} = 0$$

where the Mach number $M^2 = q^2/c^2$.

The entropy pairs we use are found from the Chaplygin functions

$$H = H_n(\mu) \, e^{\pm in\theta} \quad \text{and} \quad H = K_{\pm n}(\mu) \, e^{\pm n\theta}.$$

Here $H_n(\mu)$ is chosen to be the solution of $-\tau^2 n^2 (1 - M^2) H_n + H_{n\mu\mu} = 0$ which grows exponentially like $e^{-n\mu}$ as $\mu \longrightarrow -\infty$ and is 1 for $\mu = 0$. For $\mu > 0$, H_n behaves roughly like an Airy function but may be singular at cavitation speed ($\mu = \mu^*$).

Using the two pairs generated by $H_n e^{\pm in\theta}$ in the div-curl lemma one obtains an identity on interchanging variables appropriately. With $\cdot$ denoting differentiation,

$$0 = \iint \{ (H_n(\mu) - H_n(\mu'))(\dot{H}_n(\mu) - \dot{H}_n(\mu')) + 4 H_n(\mu) \dot{H}_n(\mu') \sin^2 \tfrac{n}{2}(\theta - \theta') \} \, dm \, dm'$$

from which we shall prove

Lemma 1: dm is a Dirac measure if dm = 0 for $\mu \geq 0$.

Proof: All terms in the integrand are negative where $\mu \neq \mu'$, $\theta \neq \theta'$ and zero for $\mu = \mu'$, $\theta = \theta'$. Hence dm = dm' = 0 for $\mu \neq \mu'$ or $\theta \neq \theta'$. Hence dm is a Dirac measure.

254

Lemma 2: If dm ≠ 0 in $\mu \leqslant 0$ then dm is a point measure.

To prove this lemma we must exclude the stagnation point where $\mu = -\infty$ and we also want to avoid the cavitation speed for technical reasons so we make the next

Assumption 2: $-\infty < \mu_1 \leqslant \mu \leqslant \mu_2 < \mu_*$ where dm ≠ 0.

A much simpler proof of this key lemma than that in [10] is given here.

Proof of Lemma: By symmetry we need only consider $\mu \geqslant \mu'$. For $\mu \geqslant 0$, $\mu' \geqslant 0$ the integrand is easily shown to be O(n). For $\mu \leqslant 0$, $\mu' \leqslant 0$ the integrand is negative. For $\mu < 0$ and $\mu' \geqslant 0$, $\mu \neq \mu'$, since $H_n \sim e^{-n\mu}$ we find the integrand $\sim$

$$-ne^{-2n\mu}(1 - o(1)) - 4ne^{-n\mu} \, O(1)$$

and hence $\longrightarrow -\infty$ like $e^{-2n\mu}$ as $n \longrightarrow \infty$. Thus dm dm' = 0 for $\mu < 0$, $\mu' \geqslant 0$, $\mu \neq \mu'$. But dm ≠ 0, hence dm' = 0 i.e. there is no measure in the supersonic region. By Lemma 1 the measure is a point measure in the subsonic region.

We now make a last assumption:

Assumption 3: For $\mu \geqslant 0$, $|\theta|$ is bounded.

Lemma 3: On $\mu = 0$, dm has support at most at one point P and the support of dm lies between the two characteristics, of the equation for H, issuing from P.

The proof uses the analogous identity for the entropy pairs generated by $K_n(\mu) \, e^{\pm n\theta}$ where K_n is decaying exponentially in $\mu > 0$, n > 0.

Lemma 4: If dm has support in $\mu \geqslant 0$ then it is a Dirac measure.

Outline of proof: If dm has no support on $\mu = 0$ Di Perna's method of proof [12] for purely hyperbolic problems can be used to prove the lemma. If it has support on $\mu = 0$, the geometric condition of Lemma 3 permits us to modify Di Perna's method. Details are in Morawetz [10].

Combining these lemmas we see that dm is a Dirac measure and we have the desired weak limits.

Proof that the entropy condition holds weakly is to be found in [10].

4. Conclusions

We have reduced the model problem of finding a potential flow past a symmetric bump to a plausible existence theorem for a third order system. In addition we have had to assume some bounds on the flow variables in order to apply a compensated compactness argument.

References

[1]   Chaplygin, S.A., On gas jets, Sci. Mem. Moscow Univ. Math.
       Phys. No. 21, 1904, pp. 1-21. Trans. NACA TM, 1963 (1944).

[2]   Tricomi, F., Sulla equatione lineari alle derivate partiali di
       secondo ordine, di tipo misto, Rendiconti Atti del Academia
       Nazionale dei Lincei, Series 5, 14 (1923), 134-247.

[3]   Busemann, A., Widerstand bei geschwindgkeiten naher der
       schallgeschwindgkeiten, Proc. Third Internat. Congr. Appl.
       Mech. 1 (1930), 282-285.

[4]   Taylor, G.I., The flow around a body moving in a compressible
       fluid, Proc. Third Internat. Congr. Appl. Mech. 1 (1930),
       263-275.

[5]   Morawetz, C.S., On the nonexistence of continuous transonic
       flows past profiles. I. Comm. Pure Appl. Math. 9 (1956),
       45-68; II. 10 (1957), 107-132; III. 11 (1958), 129-144.   See
       also 17 (1964), 357-367.

[6]   Shiffman, M., On the existence of subsonic flows of a
       compressible fluid, J. Rational Mech. Anal. 1 (1952), 605-652.

[7]   Bers, L., Existence and uniqueness of a subsonic flow past a
       given profile, Comm. Pure Appl. Math. 7 (1945), 441-504.

[8]   Bauer, F., Garabedian, P. and Korn, D., A theory of
       supercritical wing sections with computer programs and
       examples, Lecture Notes in Econom. and Math. Syst., Vol. 66,
       Springer-Verlag, Berlin and New York, 1972.   See also with
       A. Jameson, II, 108, same series and III, 105, same series.

[9]   Jameson, A., Iterative solution of transonic flow over airfoils and wings including flows at Mach 1, Comm. Pure Appl. Math. 27 (1974), 283–309.

[10]   Morawetz, Cathleen S., On a weak solution for a transonic flow problem, CPAM, 38 (1985), pp. 797–818.

[11]   Murat, F., Compacite par compensation, Ann. Scuola Norm. Sup. Pisa 5 (1978), pp. 489–507.

        Tartar, L.C., Compensated compactness and applications to partial differential equations, Nonlinear Analysis and Mechanics, Heriot–Watt Symposium IV (1979), 136–192. Research Notes in Mathematics, Pitman.

[12]   DiPerna, R.J., Convergence of approximate solutions to conservation laws, Arch. Rat. Mech. Anal. 82 (1983), pp. 27–70.

SCATTERING THEORY FOR THE WAVE EQUATION
ON A HYPERBOLIC MANIFOLD

Ralph Phillips*
Stanford University

§1. Introduction

This talk is a survey of work done jointly, mainly with Peter Lax [3,4,5] but also with Bettina Wilkott and Alex Woo [10], over the past ten years. It deals with the spectrum of the perturbed (and unperturbed) Laplace-Beltrami operator acting on automorphic functions on an n-dimensional hyperbolic space $\mathbb{H}^n$. The associated discrete subgroup Γ of motions is assumed to have the finite geometric property, but is otherwise unrestricted. This means that the fundamental domain, when derived by the polygonal method, has a finite number of sides; its volume may be finite or infinite and it can have cusps of arbitrary rank. With the obvious identifications the fundamental domain can be treated as a manifold, M.

Since signals for the wave equation propagate both along geodesics (rays) and horospheres (plane waves), we believe that the wave equation is a natural tool for the study of the Laplace-Beltrami operator and, accordingly, we have taken this approach to the problem. The principal tool in this development is the non-Euclidean version of the Radon transform which was used by Peter Lax and myself to construct an explicit translation representation for the unperturbed system. This in turn is one of the principal ingredients in our proof of the existence and completeness of the wave operators for the perturbed system. It follows from the completeness of the wave

---

*   Research partially supported by National Science Foundation under Grant DMS-85-03297 and the Ford Foundation.
   Research supported in part by NSF Grant DMS-8120790.

operators that the continuou part of the spectrum of the Laplace-Beltrami operator on M with a short range perturbation is absolutely continuous with uniform multiplicity on $(-\infty, -(\frac{n-1}{2})^2]$.

The Laplace-Beltrami operator was first studied in this context as an operator on a Hilbert space by Maass [6] in 1949. Since then this idea has been successfully exploited by many mathematicians including Roelcke [11], Patterson [7] and Elstrodt [1] who developed a spectral theory for the unperturbed Laplacian in two-dimensions. Recently Peter Perry [8], using the stationary scattering theory techniques associated with the Schroedinger wave equation, has been able to treat long range perturbations in n-dimensions on a somewhat limited class of manifolds. Results on the related problem of establishing the analytic continuation of the Eisentein series in this general setting have been obtained by S. Agmon, N. Mandouvalis, R. Melrose and P. Perry.

This talk is divided into three parts: The first and second dealing with the unperturbed operator on $\mathbb{H}^n$ and on a manifold, respectively, and the last dealing with the perturbed operator on a manifold. The first two parts consists of work done jointly with Peter Lax and the last part consists of joint work with Bettina Wiskott and Alex Woo. To simplify the exposition I shall limit myself to three dimensions.

§2. Hyperbolic 3-space

The usual model for hyperbolic 3-space is the upper half-space: { $w=(x,y)$; $x=(x_1,x_2)$, $y>0$ } with the line element

(2.1) $$ds^2 = (dx^2 + dy^2)/y^2 = g^0_{ij} dx_i dx_j \; ;$$

in the last expression we have set $y = x_3$. In this model the set of points at infinity, which we denote by B, is the union of the plane {$y=0$} and ∞. The associated Laplace-Beltrami operator is

260

(2.2) $$\Delta_0 = y^2(\partial_2^2+\partial_y^2) - y\partial_y$$

and the non-Euclidean wave equation is

(2.3) $$u_{tt} = \Delta_0 u + u \equiv L_0 u.$$

The initial value problem

(2.4) $$u(w,0) = f_1(w), \quad u_t(w,0) = f_2(w)$$

has a unique solution for all smooth initial data $f = (f_1, f_2)$.

It follows from (2.3) that the energy form defined by

(2.5) $$E_0(u) = (u_t, u_t) - (L_0 u, u)$$

is conserved for solutions of the wave equation; here $(\cdot, \cdot)$ denotes the L_2 inner products. We note that an integration by parts brings the energy into a more symmetric form:

(2.6) $$E_0(f) = \left[y^4 \left\{ |\partial_x(\tfrac{f_1}{y})|^2 + |\partial_y(\tfrac{f_1}{y})|^2 \right\} + |f_2|^2 \right] \frac{dx\,dy}{y^3}$$

which is obviously positive definite on data with compact support. By completing the space of all such initial data with respect to E_0 we obtain the Hilbert space $\mathcal{H}_0$. We denote by $U_0(t)$ the operator relating initial data in $\mathcal{H}_0$ to the solution data at time t, which is also in $\mathcal{H}_0$. $U_0(t)$ defines a strongly continuous group of unitary operators on $\mathcal{H}_0$.

The Radon transform in $\mathbb{H}^3$ is defined in terms of horospheres which are the non-Euclidean analogue of planes. The horosphere $\xi(x, \beta)$ through the point β of B and at a distance s from the 'origin' $j = (0,0,1)$ is the Euclidean sphere

(2.7)
$$|x-\beta|^2 + y^2 - y(1+|\beta|^2)e^{-s} = 0.$$

When s is negative the horosphere contains j. The symbol $<w,\beta>$ is equal to s for all points w on the horosphere $\xi(s,\beta)$. The Radon transform $\hat{u}$ of a function on $\mathbb{H}^3$ is defined as its integral over horospheres:

(2.8)
$$\hat{u}(s,\beta) = \int_{\xi(x,\beta)} u(w)\, dS,$$

dS being the non–Euclidean surface element over $\xi(s,\beta)$.

It can be shown that

(2.9)
$$e^s(L_0 u)^\wedge = \partial_s^2 e^s \hat{u}.$$

Hence taking the Radon transform of both sides of the wave equation (2.3) we see for $k = e^s \hat{u}$ that

(2.10)
$$k_{tt} - k_{ss} = (\partial_t + \partial_s)(\partial_t - \partial_s)k = 0.$$

Setting $m = k_t - k_s$ we get

(2.11)
$$m_t + m_s = 0,$$

the solution of which is of the form $m(s-t,\beta)$. This is clearly a translation representation for the solution to the wave equation. By reversing the factors in (2.10) we obtain a second translation representation. These representations lack only the property of being isometries and this is easily remedied.

We define the outgoing and incoming translation representations R_+ and R_- of data f in $\mathcal{H}_0$ as

(2.12)
$$R_\pm f = \frac{1}{2\sqrt{2\pi}} [\partial_s^2(e^s \hat{f}_1) \mp \partial_s(e^s \hat{f}_2)].$$

262

R_+ transmutes the action of $U_0(t)$ into translation:

(2.13)
$$R_+U_0(t)f = T(t)R_+f,$$

where $T(t)$ is right translation by t:

(2.14)
$$T(t)k(s, \beta) = k(s-t, \beta).$$

Similarly R_- is a left translation representation, i.e.

(2.15)
$$R_-U_0(t)f = T(-t)R_-f.$$

In addition R_+ and R_- are unitary maps of $\mathcal{H}_0$ onto $L_2(\mathbb{R} \times B)$:

(2.16)
$$E_0(f) = \|R_\pm f\|,$$

where the norm on the right is

(2.17)
$$\|R_\pm f\|^2 = \int_B \int_{-\infty}^{\infty} |R_+f(s, \beta)|^2 \, dx \, dm(\beta),$$

$$dm(\beta) = \frac{3 \, d\mu}{(1 + |\beta|^2)^2} \, .$$

It should be noted that the range of $R_\pm$ is all of $L_2(\mathbb{R} \times B)$ in contrast to the fact that the range of the Radon transform on $L_2(H^3)$ is a rather ill-defined subspace of $L_2(\mathbb{R} \times B)$.

The inverse of R_+ and R_- can be written down explicitly:

$$J_\pm k = (f_1, f_2),$$

(2.18)
$$f_1(w) = \frac{1}{2\sqrt{2}\pi} \int_B e^{<w, \beta>} k(<w, \beta>, \beta) \, dm(\beta),$$

263

$$f_2(w) = \frac{1}{2\sqrt{2}\pi}\int_B e^{<w,\,\beta>}k'(<w,\beta>,\beta)\ dm(\beta),$$

where $k'(s,\beta) = \partial_s k(s,\beta)$.

If $R_+f = 0$ for $s < 0$, then it follows from (2.13) and (2.18) that $U_0(t)f$ vanishes in the ball $B(t)$ of radius t about j for all $t > 0$. The converse is also true. We call such data outgoing and denote the class of all such data by $\mathcal{D}_+$. Similarly for incoming data f in $\mathcal{D}_-$, $R_-f = 0$ for all $s < 0$ and $U_0(-t)f$ vanishes in the ball $B(t)$ for all $t > 0$.

§3. The unperturbed system on a manifold

The metric on the manifold is inherited from $\mathbb{H}^3$ and is given as before by (2.1). Consequently the wave equation (2.3) is locally the same. However globally the situation is quite different. Although the energy form is still given by (2.5), the integration by parts used to derive (2.6) is no longer valid. The best we can do is to express E_0 locally as

(3.1) $$E_0(f) = \int_M \left[y^3(|\partial_x f_1|^2 + |\partial_y f_1|^2) - |f_1|^2 + |f_2|^2\right]\frac{dx\,dy}{y^3}$$

and it is clear from this that E_0 need no longer be positive on smooth initial data of compact support.

To get around this difficulty we construct a locally positive definite form:

(3.2) $$G_0 = E_0 + K_0,$$

where K_0 is compact with respect to G_0. Our Hilbert space $\mathcal{H}$ is then obtained as the completion with respect to G_0 of all $C_0^\infty(M)$

264

data. Roughly speaking G_0 is locally of the form (2.6). More precisely we choose a finite partition of unity $\{\varphi_i\}$, subordinate to the charts of M and define G_0 in the local coordinate system of a chart as

$$(3.3) \qquad G_j^{\varphi}(f) = \int \varphi_j \left[P^2 (| \partial_x (\tfrac{f_1}{P}) |^2 + | \partial_y (\tfrac{f_1}{P}) |^2 + | f_2 |^2 \right] \frac{dx\,dy}{y^3}$$

except on interior charts where the term

$$(3.3)' \qquad \int \varphi_j | f_1 |^2 \frac{dx\,dy}{y^3}$$

is added; here $P = y$ except on charts of cusps of intermediate rank. G_0 itself is then defined as

$$(3.4) \qquad G_0 = \Sigma \, G_j^{\varphi} \, .$$

Different partitions of unity define equivalent G_0's. Again the solution operators $U_0(t)$ form a strongly continuous group of E_0-unitary operators on $\mathcal{H}$.

The spectrum of L_0 is nonpositive except for a finite number of positive eigenvalues. We denote these eigenvalues by

$$\{ \lambda_j^2; \; j = 1,\dots,m_0 \}, \qquad \lambda_j > 0,$$

and the corresponding eigenfunctions by ς_j: $L_0 \varsigma_j = \lambda_j^2 \varsigma_j$. The functions $\exp(\pm\lambda_j t)\varsigma_j$ are solutions of the wave equation on M; we denote their initial data by

$$(3.5) \qquad p_j^{\pm} = \{ \varsigma_j, \pm\lambda_j \varsigma_j \}.$$

Obviously

(3.6) $U_0(t)p_j^+ = \exp(\lambda_j t)p_j^+$ and $U_0(t)p_j^- = \exp(-\lambda_j t)p_j^-$.

The data $p_j^{\pm}$ have finite G_0 norm and satisfy

(3.7) $E_0(p_j^+,p_k^+) = 0 = E_0(p_j^-,p_k^-)$

$E_0(p_j^+,p_k^-) = -2\lambda_j\lambda_k\delta_{jk}$

for all j,k.

Let $\mathcal{P}_0$ denoted the span of the $\{p_j^{\pm}\}$. It is clear from (3.7) that E_0 is nondegenerate on $\mathcal{P}_0$. Hence if we denote the E_0-orthogonal complement of $\mathcal{P}_0$ by $\mathcal{H}_c^0$, then every f in $\mathcal{H}$ has a unique decomposition of the form $f = g+p$, where g lies in $\mathcal{H}_c^0$ and p in $\mathcal{P}_0$. We denote the projection $f \to g$ by Q_0. Q_0 is E-orthogonal and commutes with U_0.

The energy form E_0 is nonnegative on $\mathcal{H}_c^0$. In general E_0 can have a nontrivial null space in $\mathcal{H}_c^0$. However to simplify the exposition we shall assume that E_0 is positive on $\mathcal{H}_c^0$. In that case it can be shown that E_0 and G_0 are equivalent forms on $\mathcal{H}_c^0$.

Next we define the incoming the outgoing translation representations $R_{\pm}^F$ of data f in $\mathcal{H}$. For this purpose it is convenient to think of M as a fundamental domain F contained in $\mathbb{H}^3$. Let B_F denote the union of those sides of F which lie in B and let π_j, j=1,...,N denote the parabolic fixed points for the cusps of maximal rank. Then $R_{\pm}^F$ is defined as the direct sum of components:

$$(3.8) \qquad R_{\pm}^{F}f = \sum_{j=0}^{N} \oplus (R_{\pm}^{j}f).$$

The zeroth component is simply

$$(3.9) \qquad R_{\pm}^{0}f = R_{\pm}f$$

restricted to β in B_F. Notice that for large negative s the horosphere $\xi(s,\beta)$ will have points in common with several fundamental domains. Such a horosphere will wrap itself around the manifold many times.

The rest of the components $R_{\pm}^{j}$, $j=1,...,N$, are associated with cusps of maximal rank and are defined as follows: Map the parabolic point π_j of such a cusp into ∞. Then the parabolic subgroup Γ_j of Γ leaving ∞ fixed is one of the crystallographic groups of $\mathbb{R}^2$. Let F_j be a fundamental polygon of Γ_j in $\mathbb{R}^2$; the transformed cusp is a half-cylinder of the form $F_j \times (a,\infty)$. We set

$$(3.10) \qquad R_{\pm}^{j}f(s) = \frac{1}{(2|F_j|)^{1/2}} \Big[\partial_s (e^{-s}\bar{f}_1(e^s)) - e^{-s}\bar{f}_2(e^s) \Big],$$

where

$$(3.11) \qquad \bar{f}_k(y) = \int_{F_j} f_k(w) \, dx \qquad \text{for } k = 1,2.$$

The basic properties of $R_{\pm}^{F}$ are the same as before:

(i) $R_{\pm}^{F}$ is a unitary map of $\mathscr{H}_c^0$ onto $L_2(\mathbb{R}_x B_F) \times L_2(\mathbb{R})^N$:

$$(3.12) \qquad E_0(f) = \|R_{\pm}^{F}(f)\|^2;$$

267

(ii) $R_{\pm}^F$ transmutes the action of $U_0(t)$ into translation:

(3.13)
$$R_{\pm}^F U_0(t) = T(\pm t)R_{\pm}^F f.$$

Incoming and outgoing subspaces are defined as before. Thus a solution $u(w,t)$ is called outgoing if it vanishes in the ball $B(t)$ in the manifold M, centered at j and of radius $t \geqslant 0$. Incoming solutions are defined analogously with t replaced by $-t$. We call initial data of outgoing solutions outgoing and denote the set of all such data by $\mathcal{D}_+$. Incoming data are denoted by $\mathcal{D}_-$.

If f belongs to $\mathcal{D}_+$ [or $\mathcal{D}_-$] then $R_+^F f = 0$ [$R_- f = 0$] for all $s <$ 0. Conversely if for a given f in $\mathcal{H}_c^0$, $R_{\pm}^F f = 0$ for all $s < 0$ then there exists a $d_{\pm}$ in $\mathcal{D}_{\pm}$ and a $p_{\pm}$ in the span of $\{p_j^{\mp}\}$ such that $f = d_{\pm}+p_{\mp}$. Clearly $Q_0 d_{\pm} = f$ and it can be shown that

$$\|f\|_{E_0} = \|d_{\pm}\|_{E_0} .$$

§4. The perturbed system on M

The perturbed wave equation on M is

(4.1)
$$u_{tt} = Lu$$

where

(4.2)
$$L = \frac{1}{\sqrt{g}}\partial_i\sqrt{g}g^{ij}\partial_j + q \equiv \Delta + q,$$

and Δ is the Laplace–Beltrami operator for the metric

(4.3)
$$ds^2 = g_{ij}dx_i dx_j \; .$$

We impose the following conditions on g_{ij} and q, which we state in terms of the non-Euclidean distance r from the 'center' j of the manifold and the local coordinates of the charts in M: q belongs to L_2^{loc} and g_{ij} lies in $C^{(1)}(M)$;

(4.3) $\quad \frac{1}{\mu}(g_{ij}) \leqslant (g^0_{ij}) \leqslant \mu(g_{ij}) \quad$ for some $\mu > 1$;

$$| g^{ij} - g_0^{ij} | = y^2 O(1/r^\alpha)$$

$$\left| \frac{\partial_i \sqrt{g} g^{ij}}{\sqrt{g}} - \frac{\partial_i \sqrt{g_0} g_0^{ij}}{\sqrt{g_0}} \right| = y \, O(1/r^\alpha) \quad \text{for some } \alpha > 1;$$

$$| q + \Delta P/P | = O(1/r^\beta) \quad \text{for some } \beta > 2;$$

here again P = y for all charts except those of cusps of intermediate rank. If $\alpha > 2$ then the condition on q can be stated more simply as

(4.3)'
$$| q - (\tfrac{n-1}{2})^2 | = O(1/r^\alpha).$$

We also require the unique continuation property for L.

The set-up for the perturbed system and unperturbed system on M are very similar. The perturbed energy form is

(4.4)
$$E(f) = \int_M [g^{ij}\partial_i f_1 \overline{\partial_j f_1} - q|f_1|^2 + |f_2|^2]\sqrt{g} \; dx$$

and as before E need not be positive on $C_0^\infty(M)$ data. Again, we define a G form in terms of a partition of unity $\{\varphi_j\}$. In this case the relation (3.3) becomes

(4.5)
$$G_j^{\varphi}(f) = \int \varphi_j [P^2 g^{ij}\partial_i(\tfrac{f_1}{P}) \overline{\partial_j(\tfrac{f_1}{P})} + |f_2|^2]\sqrt{g} \; dx,$$

269

and (3.2) is replaced by

(4.6)   G = E + K,

K being compact with respect to G. The forms G and G_0 are equivalent so that on completing $C_0^\infty(M)$ data with respect to G we arrive at the same Hilbert space of data $\mathcal{H}$ as before. The solution operators U(t) define a strongly continuous group of E-unitary operators on $\mathcal{H}$ with infinitesimal generator

(4.7) $A = \begin{bmatrix} 0 & I \\ L & 0 \end{bmatrix}.$

The operator L has only a finite number of positive eigenvalues. We use the corresponding eigenfunctions to construct initial data $\{q_j^{\pm}\}$ exactly as in (3.5) and these data have properties analogous to the p's in (3.6) and (3.7). Finally we denote the span of the $\{q_j^{\pm}\}$ by $\mathcal{P}$ and define $\mathcal{H}_c$ to be the E-orthogonal complement of $\mathcal{P}$ and Q to be the E-orthogonal projection of $\mathcal{H}$ on $\mathcal{H}_c$. Q commutes with the solution operators U(t). As before the energy form E is nonnegative on $\mathcal{H}_c$ and to simplify matters we shall assume that it is positive on this subspace. In this case E and G are equivalent forms on $\mathcal{H}_c$.

At this point the similarity between the two developments ends. There is no suitably endowed Radon transform for the metric (4.3) and hence we are unable to construct translation representors directly. Instead we follow the more traditional route in scattering theory and make use of wave operators to study the spectrum of the perturbed system (see [9], [12] and [13]).

The wave operators are defined on $\mathcal{H}_c^0$ to $\mathcal{H}_c$ by the relation

270

(4.8)
$$W_t = \text{st.} \lim_{t \to \pm \infty} W(t),$$

where

(4.9)
$$W(t)f = QU(-t)U_0(t)f.$$

We shall limit our discussion to the positive wave operator W_+. The existence of W_+ is proved as usual by following Cook's recipe. One uses the fact that to each f in $\mathcal{H}_c^0$ there corresponds a d_+ in $\mathcal{D}_+$ and a real number a such that $U_0(t)f \sim U_0(t-a)d_+$ for large t. This fact can also be used to prove that W_+ is an isometry.

To prove completeness we show that the range of W_+ fills out the continuous part of the spectrum of A which is

(4.10)
$$\mathcal{H}_1 = \mathcal{H}_c \ominus \text{eigenspace of A in } \mathcal{H}_c.$$

Since the spectrum of U_0 is absolutely continuous on $\mathcal{H}_c^0$, it follows that the range of W_+ is contained in $\mathcal{H}_1$ and since W_+ is an isometry its range is a closed subspace of $\mathcal{H}_1$. Hence if the range of W_+ does not fill out $\mathcal{H}_1$, there exists a nonzero f in $\mathcal{H}_1$ orthogonal to the range of W_+. Using the intertwining property of W_+ we can replace f by a time smoothed version in $\mathcal{H}_1 \cap D(A^\infty)$. We shall prove that $A^2 f = 0$ and since A has no null vectors in $\mathcal{H}_1$ it will follow that $f = 0$. Since this is contrary to our choice of f, completeness follows.

We begin by proving a local energy decay theorem. This enables us to find a sequence $\{t_n\}$, $t_n \to \infty$, such that

(4.11)
$$\|[U(t_n)f]_2\|^{r<n^2} + \sum_{j=1}^{4} \|U(t_n)A^j f\|_G^{r<n^2} < \epsilon_n,$$

271

where the ϵ_n converges to zero in a suitable fashion. We set $f_n = U(t_n)f$. Next we make use of the incoming and outgoing translation representations (for the unperturbed system on M) and define

(4.12) $$d_n = R_+^F Q_0 A f_n \quad \text{and} \quad \ell_n = R_-^F Q_0 A f_n.$$

It is easy to show that

(4.13) $$\|k_n\| + \|\ell_n\| \leqslant \text{const.}$$

independent of n. Choosing x in $C^\infty(\mathbb{R})$ so that $0 \leqslant x \leqslant 1$ and

$$x(s) = \begin{cases} 0 & \text{for } s < 0 \\ 1 & \text{for } s > 1, \end{cases}$$

we set

(4.14) $$g_n = (R_+^F)^{-1}(xk_n) \quad \text{and} \quad h_n = (R_-^F)^{-1}(x\ell_n);$$

here $(R_\pm^F)^{-1}$ denotes the inverse of the restriction of $R_\pm^F$ on $\mathcal{H}_c^0$. Thus g_n and h_n belong to $\mathcal{H}_c^0$ and by (4.13)

(4.15) $$\|g_n\|_{E_0} = \|xk_n\| \leqslant \|k_n\| \leqslant \text{const.} \quad \text{and}$$

$$\|h_n\|_{E_0} = \|x\lambda_n\| \leqslant \|\ell_n\| \leqslant \text{const.}$$

We see from (4.12) and (4.14) that

(4.16) $$E_0(Q_0 A f_n, g_n) = (k_n, xk_n) \geqslant \|xk_n\|^2,$$

$$E_0(Q_0 A f_n, h_n) = (\ell_n, x\ell_n) \geqslant \|x\ell_n\|^2.$$

272

We can prove by following an argument due to Enss [2] that

(4.17) $\lim\limits_{n\to\infty} E_0(Q_0Af_n,g_n) = 0 = \lim\limits_{n\to\infty} E_0(Q_0Af_n,h_n).$

It follows from this and (4.16) that

(4.18) $\|k_n\|^{s>1} + \|\ell_n\|^{s>1} \longrightarrow 0.$

Substracting the outgoing representation of Q_0Af_n from the incoming representation, we see from (2.12), (3.9) and (3.10) that

$\|\partial_s(e^s\hat{f}_2)\|^{s>1} \longrightarrow 0$ and

(4.20)

$\|\exp(-s)[Q_0Af_n]_2^-\|^{s>1} \longrightarrow 0.$

This is just what is needed in parts (i) and (ii) of Theorem 3.2 in [3] to prove that for a given $\epsilon > 0$ there exists a ρ_ϵ such that outside of a neighborhood of the cusps of intermediate rank

(4.21) $\overline{\lim\limits_{n\to\infty}} \|[Q_0Af_n]_2\|^{r>\rho_\epsilon} < \epsilon.$

A similar but more elaborate argument shows that this result also holds for neighborhoods of cusps of intermediate rank (see [10]). Further it is easy to show that

(4.22) $\lim\limits_{n\to\infty} \|Q_0Af_n-Af_n\|_G = 0.$

Hence (4.21) holds for Af_n in place of Q_0Af_n. Combining this with (4.11) we get

(4.23)
$$\overline{\lim_{n \to \infty}} \, \|[Af_n]_2\| < \epsilon$$

and since ϵ is arbitrary we conclude that

(4.24)
$$\lim_{n \to \infty} \|[Af_n]_2\| = 0.$$

We can replace f by Af and A^2f in the previous development to obtain

(4.24)'
$$\lim_{n \to \infty} \|[A^j f_n]_2\| = 0 \quad \text{for } j=1,2,3.$$

Now for $\{v_1, v_2\} = A^2 f_n$, we have

(4.25) $v_1 = [Af_n]_2, \quad v_2 = [A^2 f_n]_2 \quad \text{and} \quad Lv_1 = [A^3 f_n]_2$

and since

(4.26)
$$E(A^2 f_n) = (v_2, v_1) - (Lv_1, v_1)$$
$$= \|[A^2 f_n]_2\|^2 - ([A^3 f_n]_2, [Af_n]_2),$$

it follows that

(4.27)
$$\|A^2 f\|_E = \|A^2 f_n\|_E \longrightarrow 0.$$

As explained above this is enough to prove the completeness of W_+.

REFERENCES

1. Elstrodt, J., Die Resolvente zum Eigenwertproblem der automorphen Formen in der hyperbolischen Ebene, I, Math. Ann., 203 (1973) 295–330; II, Math. Zeitschr., 132 (1973) 99–134; III, Math. Ann., 208 (1974) 99–132.

2. Enss, V., Asymptotic completeness for quantum mechanical potential scattering, Comm. Math. Phys., 61 (1978) 285–291.

3. Lax, P. and Phillips, R., Translation representations for automorphic solutions of the wave equation in non-Euclidean spaces, Comm. Pure and Appl. Math. I, 37 (1984) 303–328; II, 37 (1984) 779–813; III, 38 (1985) 179–207.

4. Lax, P. and Phillips, R., Scattering theory for automorphic functions, Annals of Math. Studies, 87, Princeton Univ. Press, 1976.

5. Lax, P., and Phillips R., The asymptotic distribution of lattice points in Euclidean and non-Euclidean spaces, Jr. of Functional Analysis, 46 (1982) 280–350.

6. Maass, H., Uber eine neue Art von nichtanalytischen automorphen Functionen und die Bestimmung Dirichletscher Reihen durch Functionalgleichungen, Math. Ann., 121 (1949) 141–183.

7. Patterson, S.J., The Laplace operator on a Riemann surface, I, Compositio Math., 31 (1975) 83–107; II, Compositio Math., 32 (1976) 71–112; III, Compositio Math., 33 (1976) 227–259.

8. Perry, Peter, The Laplace operator on a hyperbolic manifold, I. Spectral and scattering theory, Jr. Funct. Anal., to appear.

9.   Phillips, R., Scattering theory for the wave equation with a short range perturbation, Indiana Univ. Math. Jr., I-31 (1982) 609-639; II-33 (1984) 831-846.

10.   Phillips, Ralph., Wiskott, Bettina and Woo, Alex, Scattering theory for the wave equation on a hyperbolic manifold, Jr. of Funct. Anal., to appear.

11.   Roelcke, W., Das eigenwertproblem der automorphen Formen in der hyperbolischen Ebene, I, Math. Ann., 167 (1966) 292-337; II, Math. Ann., 168 (1967) 261-324.

12.   Wiskott, Bettina, Scattering theory and spectral representation of short range perturbation in hyperbolic space, Dissertation, Stanford Univ., 1982.

13.   Woo, A.C., Scattering theory on real hyperbolic spaces and their compact perturbations, Dissertation, Stanford Univ., 1980.

DETERMINANTS OF LAPLACIANS ON SURFACES

By

Peter Sarnak
Department of Mathematics
Stanford University
Stanford, California 94305

In modern quantum geometry of strings and especially in the so-called Polyakov string model [P] determinants of Laplacians play a crucial role. As a result the study of this quantity and in particular its dependence on the metric, has been very productive. Our aim here is to review some recent developments in one aspect of this subject and to point out a number of unexpected relationships. Many of the results and ideas discussed here were obtained in collaboration with B. Osgood and R. Phillips and appear in the paper [O-P-S].

We begin by defining the determinant. Throughout let (M, σ) denote a compact orientable surface with a smooth metric σ. If $\partial M \neq \emptyset$ then the boundary is also taken to be smooth. We will say M is closed if $\partial M = \emptyset$. Let $\chi(M)$ denote the Euler number of M. By Δ_σ we mean the Laplace Beltrami operator for (M, σ). If M is not closed we assume that these functions vanish on ∂M. Thus in this latter case we are considering the Laplacian with Dirichlet boundary conditions. Though we do not do so we could for most of what we say deal with Laplacians on tensors. Let $0 < \lambda_1 \leq \lambda_2 \ldots$ be the non-zero eigenvalues of Δ. For a suitable orthonormal basis $u_j(x)$

$$(1) \qquad \Delta u_j + \lambda_j u_j = 0.$$

Formally we define det Δ by

$$(2) \qquad \det \Delta = \prod_{j=1}^{\infty} \lambda_j.$$

Research supported in part by NSF Grant DMS-8120790.

In order to make sense of (2) the familiar zeta function regularization is used. For Re(s) large define

(3)
$$Z(s) = Z_\Delta(s) = \sum_{j=1}^{\infty} \lambda_j^{-s}.$$

Convergence is guaranteed by Weyl's law concerning the distribution of the eigenvalues. Using Riemann's method we write

(4)
$$Z(s) = \frac{1}{\Gamma(s)} \int_0^{\infty} \left[\sum_{j=1}^{\infty} e^{-\lambda_j t} \right] t^s \frac{dt}{t}.$$

Now the standard heat kernel asymptotics [M-S] (say in the case of M being closed) asserts that

(5)
$$\Sigma \, e^{-\lambda_j t} \, u_j^2(x) \sim \frac{1}{4\pi t} - \frac{K(x)}{12\pi} + \frac{\pi}{60} K^2(x)t + \ldots$$

as $t \longrightarrow 0$. K(x) is the Gauss curvature at x. From this it is easy to see that Z(s) is meromorphic in s and in fact regular at s = 0. We have

(6)
$$Z(0) = \begin{cases} \dfrac{\chi(M)}{6} - 1 & \text{if M is closed} \\ \dfrac{\chi(M)}{6} & \text{if } \partial M \neq \emptyset. \end{cases}$$

Hence Z(0) is a topological invariant of M. However Z'(0), which formally is $- \sum_{j=1}^{\infty} \log \lambda_j$, is much more interesting. Accordingly we define

(7)
$$\det \Delta = \exp(-Z'(0)).$$

From the above considerations it is not difficult to determine the variation of $\det \Delta_\sigma$ with scale. Thus if ρ_0 is a constant then

(8)
$$\det \Delta_{\rho_0 \sigma} = \rho_0^{Z(0)} \det \Delta_\sigma.$$

When it is convenient we will refer to det Δ_σ as det M_σ.
From (7) it may appear that det Δ is an intractable quantity, as we shall see this is not at all the case. In fact det Δ can be computed, in one sense or another, in the following examples of closed surfaces of constant curvature.

(A): $(M,\sigma) = (S^2,\sigma_0)$ where σ_0 is the standard round $K \equiv 1$ metric on the sphere. As is well known the eigenvalues of Δ_{σ_0} are $n(n+1)$, $n \geqslant 0$, with multiplicity $2n + 1$. Hence

$$(9) \qquad Z(s) = \sum_{n=1}^{\infty} \frac{2n + 1}{(n(n+1))^s}.$$

From this one may show, see Vardi [VA] that

$$-Z'_{\sigma_0}(0) = \frac{1}{2} - 4\varsigma'(-1)$$

where ς is the Riemann zeta functions. Hence

$$(10) \qquad \det \Delta = e^{1/2-4\varsigma'(-1)} = 3.19531\ldots \quad .$$

(B): $(M,\sigma) = (T^2,\sigma_0)$ where T^2 is the two torus and σ_0 is a flat metric (i.e. $K \equiv 0$). Such a space may be realized as $\mathbb{C}/\Lambda$ where $\Lambda = \langle w_1,w_2\rangle$ is a lattice in $\mathbb{C}$ generated over Z by w_1,w_2. The eigenvalues of Δ_{σ_0} are $(2\pi|\gamma|)^2$ where $\gamma \in \Lambda^*$ the lattice dual to Λ. Hence

$$(11) \qquad Z_\Lambda(s) = (2\pi)^{-2s} \sum_{\gamma \in \Lambda}{}_* |\gamma|^{-2s}.$$

Choose generators u,v say for Λ^* with $y = \mathrm{Im}(v/u) > 0$ then

$$Z(s) = (2\pi)^{-2s} \sum_{m,n}{}' \frac{1}{|mu + nv|^{2s}}$$

$$= (2\pi)^{-2s} |u|^{-2s} \sum_{m,n}{}' \frac{1}{|m + nz|^{2s}}$$

where $z = v/u \in H = \{z \mid \mathrm{Im}\, z > 0\}$. If we normalize σ_0 so that it has area 1 then $|u|^2 = y^{-1}$ and

(12) $$Z(s) = (2\pi)^{-s} E(z,s)$$

$E(z,s)$ being the Eisenstein series [SI]

(13) $$E(z,s) = \sum_{m,n}{}' \frac{y^s}{|mz + n|^{2s}}.$$

The function $Z_\Lambda(s)$ depends on the lattice Λ but not on the choice of basis for Λ^* hence

(14) $$E(Tz,s) = E(z,s)$$

for any $T \in SL(2,Z)$. Using Kronecker's first limit formula [SI], $\dfrac{\partial E(z,s)}{\partial s}\Big|_{s=0}$ may be computed explicitly in terms of the Dedekind Eta function;

(15) $$\eta(z) = e^{\pi iz/12} \prod_{n=1}^{\infty} (1 - e^{2\pi inz}).$$

The result is

(16) $$\det \Delta_z = y |\eta(z)|^4.$$

This function is $SL(2,Z)$ invariant according to (14). Another important property is that it satisfies

(17)
$$y^2 \Delta(\log \det \Delta_z) = -1 \text{ or}$$
$$\Delta_H(\log \det \Delta_z) = -1$$

where Δ_H is the Laplacian for the Lobachevsky-Bolayi upper half plane. (17) is a simple consequence of $\eta(z)$ being holomorphic in z. The eta function is well understood and easily computed with the aid of (15) and so the same is true for $\det \Delta_z$.

(C): $(M,\sigma) = (M,\sigma_0)$ where M is a closed surface of genus $g \geq 2$

and σ_0 is a hyperbolic metric (i.e. $K \equiv -1$). One approach in this case is to express det Δ in terms of the apparently more tractable (and certainly more computable) Selberg zeta function for (M,σ_0). This function is defined in terms of the lengths of the set P of primitive closed geodesics on M.

$$(18) \qquad Z_{\sigma_0}(s) = \prod_{\gamma \in P} \prod_{k=0}^{\infty} \left[1 - e^{-\ell_\gamma (s+k)} \right]$$

where ℓ_γ is the length of γ. See Selberg [SE]. From the Selberg Trace Formula one can show that $Z(s)$ is entire of order 2 and has zeros at $\frac{1}{2} + ir_j$ where $r_j^2 + \frac{1}{4} = \lambda_j$, as well as trivial zeros at $s = -k$, $k \geqslant 1$ with multiplicity $(2g-2)(2k+1)$ see [SE]. In particular since $\lambda_0 = 0$ we see that $s = 1$ is a simple zero of $Z(s)$.

In [D-P], [K], [VO], [F], [SA] it is shown that

$$(19) \qquad \det \Delta_{\sigma_0} = Z'_{\sigma_0}(1) \exp(g(4\varsigma'(-1) - \frac{1}{2} + \log 2\pi)).$$

The reason for the relation (19) is quite simple and a proof of (19) along the following lines is carried out in [SA] and [VO]. Instead of considering det Δ directly consider the function

$$(20) \qquad B(s) = \det(\Delta + s(s-1))$$

$B(s)$ is defined through the regularization in (3). It is entire in s and vanishes exactly for those s for which $s(s-1) = \lambda_j$, i.e. $s = \frac{1}{2} + ir_j$. Hence besides the trivial zeros of $Z(s)$, which may be taken into account by the double gamma function $\Gamma_2(s)$, see [BA], $B(s)$ and $Z(s)$ are entire and have the same zeros. It is then not surprising that one can show that the two functions agree up to a couple of parameters which in turn may be evaluated by letting $s \longrightarrow \infty$ and using Stirling's formula. The result is [SA]

$$(21) \qquad \det(\Delta + s(s-1)) = Z(s) \left[e^{E-s(s-1)} \frac{(\Gamma_2(s))^2 (2\pi)^2}{\Gamma(s)} \right]^{2g-2}$$

281

where $E = -\frac{1}{4} - \frac{1}{2} \log 2\pi + 2\varsigma'(-1)$. (19) then follows from (21) by specialization.

Formula (19) is a pretty one but suffers from the drawback that we do not know much about the function $Z'_{\sigma_0}(1)$ as a function of the hyperbolic surface. That is to say as a function on the space of moduli of curves of genus g. As is well known the space of moduli M_g is a 3g - 3 complex analytic space. The problem is to understand $\det \Delta_\tau$ for $\tau \in M_g$. In a remarkable announcement [B-K] Belavin and Knizhnik show that the modular function $\log \det \Delta_\tau$ satisfies an equation analogous to (17) w.r.t. the Weil Petterson metric. They show that the second variation in τ and $\bar{\tau}$ of $\log \det \Delta_\tau$ is given by an integral over M of local quantities. This is a most promising attack and opens the way for computing $\det \Delta_\tau$ in terms of Riemann theta functions [MA1]. This aspect of the study of $\det \Delta$ is an exciting and explosive area of research and we refer to [MA2] for recent developments.

The above examples are all constant curvature closed surfaces and are examples of what we will call uniform metrics. If $\partial M \neq \emptyset$ the notion of a constant curvature metric is more ambiguous. We will call M uniform if either

(I)   M has constant curvature and ∂M is of zero geodesic curvature (i.e. is a union of disjoint closed geodesics).

(II)   M is flat (i.e. $K \equiv 0$) and ∂M is of constant geodesic curvature.

These will be referred to as uniform metrics of type (I) and (II) respectively. Examples of uniform metrics of type I are a hemisphere, a cylinder and a 'pair of pants' [T]. The unit disk with its planar metric, the cylinder and a flat torus with some number of disjoint circles of equal radii removed are simple examples of type II uniform surfaces. The type I and II uniform surfaces should be thought of as dual to each other.

The importance of uniform metrics in the theory of

282

determinants stems from the variational formula of Polyakov [P] which we now describe. Let σ_0 be a fixed metric on **M**. The conformal class of this metric is the family of metrics $\delta = \rho \sigma_0$ where $\rho > 0$ and is in $C^\infty(M)$. We write $\rho = e^{2\phi}$ with $\phi \in C^\infty(M)$. One checks that

(22)
$$
\begin{cases}
dA = e^{2\phi} dA_0 \\
\Delta = e^{-2\phi} \Delta_0 \\
K = e^{-2\phi}(-\Delta_0 \phi + K_0)
\end{cases}
$$

where dA, Δ and K are the area element, Laplacian and curvature for the σ metric whild dA_0, Δ_0 and K_0 are the corresponding quantities associated with σ_0. Consider first the case of a closed surface **M**. Polyakov's variational formula reads

(23)
$$
\delta(\log \det \Delta_\phi) = -\frac{1}{6\pi} \int_M (\delta\phi)(-\Delta_0\phi) dA_0
$$
$$
-\frac{1}{6\pi} \int_M K_0(\delta\phi) dA_0 + \delta(\log A).
$$

The derivation of (23) is short. Firstly we write (4) as

$$
Z(s) = \frac{1}{\Gamma(s)} \int_0^\infty TR'(e^{\Delta t}) t^s \frac{dt}{t}
$$

TR' denoting trace after projecting out $\lambda_0 = 0$. Then

$$
\delta Z(s) = \frac{1}{\Gamma(s)} \int_0^\infty TR'(\delta \Delta e^{\Delta t}) t^s dt.
$$

From (22)

$$
\delta\Delta = -2e^{-2\phi}\delta\phi\Delta_0 = -2\delta\phi\Delta.
$$

283

Hence

$$
\delta Z(s) = \frac{1}{\Gamma(s)} \int_0^\infty TR'(-2\delta\phi\Delta e^{\Delta t})t^s dt
$$

$$
= -\frac{1}{\Gamma(s)} \int_0^\infty \frac{d}{dt} TR(-2\delta\phi(e^{\Delta t} - \frac{1}{A}))t^s \frac{dt}{t}
$$

$$
= \frac{2s}{\Gamma(s)} \int_0^\infty TR((\delta\phi)(e^{\Delta t} - \frac{1}{A}))t^s \frac{dt}{t} .
$$

The term $\frac{s}{\Gamma(s)}$ has a double zero at $s = 0$ and so what survives after differentiating and setting $s = 0$ is simply the residue at 0 of the integral. This is easily calculated from (5) and results in

$$
\delta Z'(s)\big|_{s=0} = 2 \int_M (\delta\phi)(\frac{K(x)}{12\pi} - \frac{1}{A})e^{2\phi}dA_0
$$

which leads to (23). (23) may be integrated to get

(24)
$$
\log \det \Delta_\phi = -\frac{1}{6\pi} \{\frac{1}{2} \int_M |\nabla_0\phi|^2 dA_0 + \int_M K_0\phi dA_0\}
$$
$$
+ \log A + C
$$

where C is independent of ϕ.

In the $\partial M \neq \phi$ case a similar argument see [A] leads to

(25)
$$
\log \det \Delta_\phi = -\frac{1}{6\pi} \{\frac{1}{2} \int_M |\nabla_0\phi|^2 dA_0 + \int_M K_0\phi dA_0
$$
$$
+ \int_{\partial M} k_0\phi ds_0\} - \frac{1}{4\pi} \int_{\partial M} kds + C.
$$

Here k_0 is the geodesic curvature of ∂M w.r.t. σ_0. The second to last term may be expressed as

(26)
$$\int_{\partial M} k ds = \int_{\partial M} k_0 ds_0 - \int_{\partial M} \partial_n \phi ds_0$$

∂_n being the unit outer normal derivative along ∂M in the σ_0 metric.

The relation (25) displays clearly the dependence of det Δ on the conformal factor $e^{2\phi}$. For the purpose of extremizing, which is the next thing we would like to consider, we need to impose some conditions on the metrics considered. We shall assume the two dual conditions;

CI $\quad \int_{\partial M} k ds \geq 0$ or equivalently $\int_M KdA \leq 2\pi \chi(M)$

CII $\quad \int_{\partial M} k ds \geq 2\pi \chi(M)$ or $\int_M KdA \leq 0.$

The importance of the uniform metric lies in the following theorem which states that under suitable constraints the uniform metric extremizes det Δ in a given conformal class.

Theorem 1 [O-P-S]:

(a)   *If* M *is closed then of all metrics in a given conformal class and of a given area the uniform metric maximizes* det Δ.

(b)   *If* $\partial M \neq \phi$ *then in a given conformal class of metrics all of given area and satisfying* CI *the uniform metric of type* I *has maximum determinant.*

(c)   *If* $\partial M \neq \phi$ *then in a given conformal class of metrics all of given boundary length and satisfying* CII *the uniform metric of type* II *has maximum determinant.*

Remarks:

(i)   Part of Theorem 1 is the assertion that in each conformal class of metrics there is exactly one uniform metric (up to scale and where isometric surfaces are of course considered as one and the same). When M is closed this is the classical uniformization theorem.

(ii)   When M is simply connected parts (a) and (c) are equivalent to the following inequalities (via (25)).

(A):   For $u \in C^1(S^2)$

$$\log \int_{S^2} e^{2\phi} \frac{dA_0}{4\pi} \leqslant \int_{S^2} |\nabla_0 \phi|^2 \frac{dA_0}{4\pi} + 2 \int_{S^2} \phi \frac{dA_0}{4\pi}$$

where σ_0 is the standard metric on S^2. Furthermore equality holds in (A) iff $\phi = \log|T'(z)| + \beta$ where $T: S^2 \longrightarrow S^2$ is a Moebius transformation and $\beta \in \mathbb{R}$.

(B):   For $u \in C^1(\bar{D})$ where $D = \{z \mid |z| < 1\}$ and σ_0 is the planar metric

$$\log \int_{\partial D} e^u \frac{ds_0}{2\pi} \leqslant \frac{1}{4} \int_D |\nabla u|^2 \frac{dxdy}{\pi} + \int_{\partial D} u \frac{ds_0}{2\pi}$$

with equality iff $u = \log|T'(z)| + \beta$ where $T: D \longrightarrow D$ is a Moebius transformation and $\beta \in \mathbb{R}$.

Remarkably inequality (B) is essentially the first of the Milin–Lebedev inequalities [DU] which are crucial in the solution of the Bieberbach conjecture [D]. Indeed the above gives a geometric interpretation of this inequality, it being equivalent to the fact that the determinant of a simply connected plane domain of given boundary length is maximized precisely for a circular domain. Inequality (A) is a sharpening of an inequality of Moser [MO] first derived by Onofri [O] using an inequality of Aubin. A geometric and uniform method of deriving both (A) and (B) is given in [O-P-S].

As a corolary to Theorem 1 and to examples A and B

mentioned at the beginning we have

<u>Corollary</u> [O-P-S]:

(a) $$\det \Delta_\sigma \leqslant e^{1/2-4\zeta'(-1)} = 3.19531...$$

for all metrics σ on S^2 of area 4π with equality iff $\sigma = \sigma_0$ the round metric.

(b) $$\det \Delta_\sigma \leqslant \frac{\sqrt{3}}{2} \left| \eta(\frac{1}{2} + \frac{i\sqrt{3}}{2}) \right|^4 = 0.35575...$$

for all metrics σ on T^2 of area 1 with equality iff σ corresponds to the flat metric on $\mathbb{C}/\Lambda$ where $\Lambda = \langle 1, e^{2\pi i/3} \rangle$.

A natural generalization of this corollary would be to give a similar result for surfaces of genus $g \geqslant 2$. More precisely determine the metric or metics (on such a surface) of area $-2\pi\chi(M)$ for which $\det \Delta$ is a maximum. By Theorem 1 such a metric is hyperbolic and so this problem reduces to extremizing $\log \det \Delta_\tau$ over moduli space.

An interesting case not dealt with by Theorem 1 is that of maximizing $\det \Delta$ over all plane domains (simply connected) of fixed area.

<u>Proposition 1</u>: *Of all simply connected plane domains of given area the circular domain has maximal determinant.*

Note that if we do not impose the planar condition in the Proposition then by Theorem 1 the hemisphere (i.e. type I uniformizer) would give a largest determinant.

<u>Proof of Proposition 1</u>: Let $F: D \longrightarrow \Omega$ be a conformal map of the unit disk onto a domain Ω. The space $(\Omega, |dz|^2)$ is isometric to $(D, |F'(z)|^2 |dz|^2)$. Write $|F'(z)|^2 = e^{2\phi}$. Then ϕ is harmonic and

287

in view of (2.5) our problem reduces to showing that

(27)
$$G(\phi) = \frac{1}{2\pi} \int_D |\nabla\phi|^2 dxdy + \frac{1}{\pi} \int_{\partial D} \phi d\theta$$
$$- \log \int_D e^{2\phi} \frac{dxdy}{\pi} \geq 0$$

for any harmonic function ϕ, with equality iff $\phi = \log|T'(z)| + \beta$ where $T: D \longrightarrow D$ is Moebius. Since $G(\phi+\alpha) = G(\phi)$ for $\alpha \in \mathbb{R}$ we may assume that $\int_{\partial D} \phi d\theta = 0$. Let $f(z) = \phi + i\psi$ where ψ is a harmonic conjugate of ϕ and $\psi(0) = 0$. Then $f(0) = 0$ and (27) becomes

(28)
$$\frac{1}{2\pi} \int_D |f'(z)|^2 dxdy - \log \int_D |e^{f(z)}|^2 \frac{dxdy}{\pi} \geq 0.$$

Expand $f(z)$ in a power series

(29)
$$\begin{cases} f(z) = \sum_{n=1}^{\infty} a_n z^n & \text{and} \\ g(z) = e^{f(z)} = \sum_{n=0}^{\infty} \beta_n z^n. \end{cases}$$

After some manipulation (28) becomes

(30)
$$\log \sum_{n=0}^{\infty} \frac{|\beta_n|^2}{n+1} \leq \frac{1}{2} \sum_{n=1}^{\infty} n|a_n|^2$$

(30) is an interesting variation of the Milin-Lebedev inequality which we prove as follows. Differentiating $g(z) = e^{f(z)}$ gives $g'(z) = e^{f(z)}f'(z)$ which on equating coefficients leads to

(31)
$$\beta_0 = 1 \quad \text{and for} \quad n \geq 1$$
$$\beta_n = \frac{1}{n} \sum_{j=0}^{n-1} (n-j)a_{n-j}\beta_j.$$

Let $\gamma_n = \dfrac{\beta_n}{\sqrt{n+1}}$. Then $\gamma_0 = 1$ and (31) becomes

(32)
$$\gamma_n = \frac{1}{n\sqrt{n+1}} \sum_{j=0}^{n-1} (n-j)\sqrt{j+1}\, \alpha_{n-j}\gamma_j.$$

It follows that

$$|\gamma_n|^2 \le \frac{1}{n^2(n+1)} \frac{n(n+1)}{2} \sum_{j=0}^{n-1} (n-j)^2 |\alpha_{n-j}|^2 |\gamma_j|^2$$

or

(33)
$$|\gamma_n|^2 \le \frac{1}{2n} \sum_{j=0}^{n-1} (n-j)a_{n-j}|\gamma_j|^2$$

where

$$a_n = n|\alpha_n|^2.$$

Define b_n by;

$$b_0 = 1$$
$$b_n = \frac{1}{2n} \sum_{j=0}^{n-1} (n-j)a_{n-j}b_j \qquad \text{for} \qquad n \ge 1.$$

It is apparent that then

(34)
$$\sum_{n=0}^{\infty} b_n z^n = e^{1/2 \sum_{n=1}^{\infty} a_n z^n}$$

From (33) one sees inductively that

$$|\gamma_n|^2 \le b_n.$$

Hence

$$\sum_{n=0}^{\infty} |\gamma_n|^2 \le \sum_{n=0}^{\infty} b_n = e^{1/2 \sum_{n=1}^{\infty} a_n}$$

289

which proves (30). The case of equality in the above analysis is easily examined and leads to (27).

When M is not simply connected, the extremal problem of Theorem 1 is easier to handle since in that case one can reformulate the problem in terms of a convex functional. More precisely let M be closed then from (8) and (24) it is natural to introduce the scale invariant 'log det'

$$(35) \quad F_0(\phi) = \frac{1}{2} \int_M |\nabla_0 \phi|^2 dA_0 + \int_M K_0 \phi dA_0 - \pi \chi(M) \log \int_M e^{2\phi} dA_0$$

σ_0 is an arbitrary fixed metric on M. Similarly, say for example for the type II surface, the scale invariant function is

$$(36) \quad F_2(\phi) = \frac{1}{2} \int_M |\nabla_0 \phi|^2 dA_0 + \int_{\partial M} k_0 \phi ds_0 - 2\pi \chi(m) \log \int_{\partial M} e^{\phi} ds_0.$$

This time the background metric σ_0 is assumed to be flat in the interior which is no loss of generality. In minimizing F_2 we may as well assume ϕ is harmonic and then we may restrict to the values of ϕ on ∂M.

$$(37) \quad F_2(\phi) = \frac{1}{2} \int_{\partial M} \phi \partial_n \phi ds_0 + \int_{\partial M} k_0 \phi ds_0 - 2\pi \chi(m) \log \int_{\partial M} e^{\phi} ds_0$$

∂_n is the unit outer normal derivative of the harmonic extension of ϕ. This is a non-local operator and makes matters technically more difficult in what follows. By Theorem 1 the uniformizer is the unique global minimum of (35) and (37) respectively. It is natural therefore to use the gradient vector field defined by F (i.e. 'log det Δ_ϕ') to construct a flow which should in the limit converge to the uniformizer. In particular such a flow would give a constructive method of uniformizing metrics.

In [O-P-S] this approach is carried out. We considered the flow

290

(38)
$$\frac{\partial \phi}{\partial t} = -\text{grad } F_j(\phi), \qquad j = 0,1.$$

An important point here is that in order to define grad F, an inner product on the tangent space at ϕ, must be chosen so that

(39)
$$\delta F(\phi)(h) = \langle \text{grad } F(\phi), h \rangle.$$

If M is closed and one uses the inner product

(40)
$$\langle h_1, h_2 \rangle_\phi = \int_M h_1 h_2 dA$$

where $dA = e^{2\phi} dA_0$ is the area element of $\sigma = e^{2\phi} \sigma_0$, then (38) becomes

(41)
$$\frac{\partial \phi}{\partial t} = \bar{K} - K$$

where $\bar{K} = \frac{1}{A} \int_M KdA$ is the average curvature. This is the flow

introduced by Hamilton [H]. It is geometrically natural and it preserves area. Clearly a steady state of (41) is the uniformizer. Technically we found it more convenient to work with the fixed inner product on M given by σ_0

(42)
$$\langle h_1, h_2 \rangle_\phi = \int_M h_1 h_2 dA_0$$

(38) with this choice becomes the nonlinear evolution equation

(43)
$$\frac{\partial \phi}{\partial t} = \Delta_0 \phi - K_0 + \frac{2\pi \chi(M) e^{2\phi}}{\int_M e^{2\phi} dA_0}.$$

Theorem 2 [O-P-S]: *For $\psi_0 \in C^\infty(M)$, equation (43) has a unique smooth solution $\psi(x,t)$ for all $t \geq 0$ such that $\psi(x,0) = \psi_0(x)$. Furthermore $\psi(x,t) \longrightarrow \psi_\infty(x)$ in the*

Sobolev k norm for all k ⩾ 0 as t ⟶ ∞, where $e^{2\psi_\infty}\sigma_0$ *is uniform.*

Similarly for F_2 in (38). In this case we have the evolution equation

(44)
$$\frac{\partial\phi}{\partial t} = \partial_n\phi - k_0 + \frac{2\pi\chi(M)e^\phi}{\int_{\partial M} e^\phi ds_0}.$$

Theorem 2′ **[O–P–S]:** *For $\psi_0 \in W^1(\partial M)$ (i.e. Sobolev 1 space) (44) has a unique global solution $\psi(x,t)$, t ⩾ 0 such that $\psi(x,0) = \psi_0$. Furthermore $\psi(x,t) \longrightarrow \psi_\infty(x)$ in the W^1 norm as t ⟶ ∞ and $e^{2\psi_\infty}\sigma_0$ is uniform.*

A similar result holds for the type I uniformizers. The proofs of Theorems 2 and 2′ are somewhat involved and 2′ is technically tricky. The details may be found in Section 3 of [O–P–S]. The above results should be useful in numerical uniformization problems.

Finally we remark that the quantity det Δ is rather useful in the inverse type problem 'Can you hear the shape of a drum'. The eigenvalues determine det Δ and hence (24) or (25) as the case may be. One may use this to prove compactness of isospectral sets and to give a number of examples of spectrally determined domains. Again we refer the reader to [O–P–S] for a description and proofs of these results.

In conclusion it appears that even without its relation to Physics, the determinant of the Laplacian is an important isometric invariant for surfaces.

References

[A]   Alvarez, O., "Theory of strings with boundaries," Nuclear
         Phys. B.216 (1983) 125–184.

[BA]   Barnes, E.W., "The theory of the G-function," Quart.
         Jnl. of Math. Vol. 31, 264–314 (1900).

[BE]   Belavin, A., and V. Knizitnik, Phys. Lett. B.168 (1986)
         201.

[D]   DeBranges, L., "A proof of the Bieberbach conjecture,"
         Acta Math. 154 (1985) 137–152.

[D-P]   D'Hoker, E., and D.H. Phong, Comm. Math. Physics 105
         (1986) 537.

[DU]   Duren, P., "Univalent functions," Springer Verlag,
         New York 1983.

[F]   Fried, D., Invent. Math. 84 (1986) 523–540.

[H]   Hamilton, R., "Three manifolds with positive Ricci
         curvature," J. Diff. Geom. 17 (1982) 255–306.

[K]   Kierlanczyk, J.M., "Determinants of Laplacians," Ph.D.
         thesis M.I.T. 1986.

[MA1]   Manin, Yu., "The partition function of the Polyakov
         string can be expressed in terms of theta functions,"
         Phys. Letters B, Vol. 172, No. 2, 184–186 (1986).

[MA2]   Manin, Yu., International Congress talk 1986.

[M-S]   McKean, H., and I. Singer, "Curvature and eigenvalues of the Laplacian," J. Diff. Geom. Vol. 1, No. 1 (1967) 43-69.

[MO]   Moser, J., "A sharp form of an inequality by N. Trudinger," Indiana Math. J. 20 (1971) 1077-1092.

[ON]   Onofri, E., "On the positivity of the effective action in a theory of random surface," Comm. Math. Phys. 86 (1982) 321-326.

[O-P-S]   Osgood, B., R. Phillips, and P. Sarnak, "Extrema of determinants of Laplacians," preprint 1986.

[P]   Polyakov, A.M., Phys. Lett. B.103 (1981) 207-211.

[SA]   Sarnak, P., "Determinants of Laplacians," preprint 1986.

[SE]   Selberg, A., Jnl. Ind. Math. Soc. 20 (1956), pp. 47-87.

[SI]   Siegel, C., "Lectures on advanced analytic number theory," Tata Inst. (1961).

[T]   Thurston, W., "Geometry and topology of three manifolds," Princeton 1979.

[VA]   Vardi, I., "Determinants of Laplacians and multiple gamma functions," preprint 1986.

[VO]   Voros, A., "Spectral functions, special functions and the Selberg zeta function," preprint 1986.

THE SMALL DISPERSION LIMIT OF THE
KORTEWEG-DE VRIES EQUATION

Stephanos Venakides[*]
Stanford University

To Peter Lax on his 60th birthday.

$$\nu\alpha \ \tau\alpha \ \epsilon\kappa\alpha\tau o\sigma\tau\eta\sigma\eta s$$

_____

Research supported in part by NSF Grant DMS-8120790.

[*]Present address: Duke University, Mathematics, Durham, NC 27706.

1.   Introduction

There are many physical systems which display shocks i.e. regions in space where the solution develops extremely large slopes. In general, such systems are too complicated to be treated by exact calculation and their properties are best studied through the proof of general theorems.   A model of the formation and propagation of dispersive shocks in one space dimension, in which explicit calculation is possible, is given by the initial value problem for the Korteweg-de Vries equation:

(1.1a) $$u_t - 6uu_x + \epsilon^2 u_{xxx} = 0$$

(1.1b) $$u(x,o,\epsilon) = -v(x)$$

in the limit $\epsilon \longrightarrow 0$.

When $\epsilon = 0$, a simple calculation using the method of characteristics demonstrates that for almost all initial data the solution of the evolution equation (1.1a) develops a shock in finite time.   When $\epsilon \neq 0$, the dispersive third derivative term in (1.1a) becomes significant as soon as slopes of order $1/\epsilon$ are reached.   An actual shock discontinuity is averted and a region where the solution is rapidly oscillatory appears.   The oscillations have frequencies and wave numbers of order $1/\epsilon$, and speeds of order 1.

The width of the oscillatory regions has order O(1) and grows with time.

This phenomenon of weak dispersion is in sharp contrast to the--better known--effects produced by weak dissipation in Burger's equation:

(1.2) $$u_t - 6uu_x - \epsilon^2 u_{xx} = 0.$$

related to the kdV equation by the fact that both are singular perturbations of the equation:

(1.3) $$u_t - 6uu_x = 0.$$

It is well known that when ε is small, but nonzero, the large slopes which appear in the solution of Burger's equation are restricted within thin boundary layers which travel with speeds of order 1. The solution is non-oscillatory and, as ε → 0 it converges strongly to a function having jump discontinuities at the shock fronts. A consequence of the above strong convergence is that the limit of the solution of the initial value problem for (1.2) as ε → 0 satisfies equation (1.3) in a weak sense.

Contrary to this, in the case of small dispersion, the oscillatory nature of the solution of equation (1.1) precludes strong convergence. Lax and Levermore show in a beautiful calculation [12] that the limit of the solution of (1.1) as ε → 0 exists in a weak sense by actually computing it. I later extended their result to a more general class of initial data including periodic ones and I developed a technique which allows the calculation of the waveform in the oscillatory regions [17,18,19]. The result is that locally the oscillation consists of a periodic or N-phase quasiperiodic soliton solution of the kdV equation (see Sec. 2) with wave parameters (i.e. amplitude, wave numbers and frequencies) varying slowly with x and t. Such solutions are referred to as modulated N-phase solitons and the slowly varying wave parameters are called modulation parameters. The formation of a shock generates an additional oscillatory phase and locally introduces two new slowly varying parameters: a wave number and a frequency. The generation and evolution of the modulation parameters is governed by an--in principal well posed--first order initial value problem.

An alternative approach is to *postulate* that the solution of equation (1.1) is locally a modulated N-phase soliton (where N = 0,1,2,... depends on x and t) and apply modulation theory to derive a set of equations which describe the evolution of the modulation parameters. In any region of the (x,t) space in which the number N of phases present in the solution is constant, the 2N + 1 modulation equations, derived by Forest, Flaschka and McLaughlin [7] are essentially identical to the equations derived by Lax and Levermore. On the other hand at the shock fronts, where a new phase (and hence a new pair of modulation parameters) is generated, the calculation via the small dispersion limit shows how to connect the

297

two systems of modulation equations on either side of the shock. I will discuss this in more detail after I have described the completely integrable character of the kdV equation and the nature of its soliton solutions.

2.   Complete Integrability of the kdV Equation

The Korteweg-de Vries equation

(2.1) $$u_t - 6uu_x + u_{xxx} = 0$$

was first solved explicitly by the inverse scattering method in a landmark paper by Gardner, Greene, Kruskal and Miura [8].

The method of the inverse scattering relies on the strong connection between the Korteweg-de Vries equation (2.1) on the one hand and the Schrödinger operator on the other. In more precise terms the spectrum of the Schrödinger operator $\mathscr{L} = \dfrac{d^2}{dx^2} - u(x,t)$ is independent of t when u(x,t) satisfies the kdV equation.

The deeper reason for this connection was brought to light by Peter Lax [11] through his discovery of the Lax pair in 1968. The Lax pair consists of two operators $\mathscr{L}$ and B parametrized by a real variable t and having the following properties:

(a)   B(t) generates a one parameter family of unitary operators U(t):

(2.2) $$\frac{d}{dt} U = B(t)U, \quad U(0) = \text{Identity}$$

This requires B(t) to be antisymmetric.

(b)   The pair satisfies the operator equation:

(2.3) $$\frac{d}{dt} \mathscr{L}(t) = [B(t),\mathscr{L}(t)]$$

where the right hand side is the commutator of B and $\mathscr{L}$. One verifies by direct differentiation that equations (2.2) and (2.3) imply:

(2.4) $$\frac{d}{dt} \{U^{-1}(t)\mathscr{L}(t)U(t)\} = 0.$$

i.e. for all t the operator $\mathscr{L}(t)$ is unitarily equivalent to $\mathscr{L}(0)$. This explains the invariance of the spectrum. Lax now chooses

(2.5) $\mathcal{L} = \dfrac{d^2}{dx^2} - u(x,t).$ $B = -4\dfrac{d^3}{dx^3} + 3u\dfrac{d}{dx} + 3\dfrac{d}{dx}u.$

B clearly satisfies the requirement (a). Substituting $\mathcal{L}$ and B in (2.3) we see that all terms containing $\dfrac{d}{dx}$ cancel and (2.3) is reduced to the scalar equation $u_t - 6uu_x + u_{xxx} = 0$ i.e. the kdV equation. Thus the pair (2.5) is a Lax pair if and only if u(x,t) satisfies the kdV equation. The connection between the Schrödinger operator $\mathcal{L}(t)$ and the kdV equation is now established.

The existence of a Lax pair formulation of the evolution of a system is utilized in integrating the system. The idea is to define a set of spectral data of the operator $\mathcal{L}(t)$ which are sufficient to determine this operator uniquely. We call these data S(t). Their evolution in t depends on the operator B(t) and is usually straightforward to resolve. The solution of the integrable system (let us say of the kdV equation) proceeds then according to the diagram:

$$u(x,0) \xrightarrow{\quad kdV \quad} u(x,t)$$

```
        | direct  spectral        | inverse  spectral
        | problem                 | problem
 S(0) ──────evolution of─────→  S(t)
           spectral data
```

The spectral data include the spectrum of $\mathcal{L}(t)$, which is independent of t, and some other piece of information which evolves with t. I will survey--without giving any proof--the solution of the kdV equation corresponding to types of initial data for which the above procedure has been carried out. Most of the work lies in the inverse problem.

(a)   *Initial data v(x) decaying fast enough for the*

integral $\displaystyle\int_{-\infty}^{\infty} (|x| + 1) |v(x)| dx$ *to be finite:*

In his solution of the inverse problem in this case, Fadeev [6; see also 3] recovers the potential u(x) of the Schrödinger operator $\mathcal{L} = \dfrac{d^2}{dx^2} - u$ from the $L_2(-\infty,\infty)$ spectrum of $\mathcal{L}$ and from the asymptotic behavior of the corresponding eigenfunctions as $x \longrightarrow \pm\infty$. The spectrum of $\mathcal{L}$ consists of the negative half-axis plus

300

at most finitely many discrete eigenvalues $n_1^2, n_2^2, ..., n_N^2$. The eigenfunction corresponding to the point $-k^2$ of the continuous spectrum having form:

(2.6a)
$$f(x,k) \sim T(k)e^{-ikx} \text{ as } x \longrightarrow -\infty$$
$$\sim e^{-ikx} + R(k)e^{ikx} \text{ as } x \longrightarrow \infty$$

has asymptotic behavior as $x \longrightarrow \pm\infty$ determined by the *transmission coefficient* $T(k)$ and by the *reflection coefficient* $R(k)$. This terminology comes from quantum mechanics, where $f(x,k)$ is the wavefunction corresponding to the scattering by the potential of a particle incident from the right. The discrete eigenvalues n_j^2 correspond to bound states i.e. energy states for which the wavefunction can be localized in space. The normalized eigenfunction $f_j(x)$ corresponding to the eigenvalue n_j^2 satisfies:

(2.6b)
$$\|f_j\|_{L_2} = 1, \quad f_j(x) \sim c_j e^{-n_j x}, \quad j = 1,...,N$$

and its asymptotic behavior is determined by the *norming constant* c_j. The transmission coefficient can be derived from the reflection coefficient; it is not surprising that it does not appear in the formalism of the inverse problem.

To recover $u(x)$ from the spectral data one first constructs the function $g(\rho)$ from the spectral data:

(2.7a)
$$g(\rho) = \frac{1}{2\pi} \int_{-\infty}^{\infty} R(k)e^{i\rho k}dk + \sum_{i=1}^{N} c_j^2 e^{-n_j \rho}.$$

One then solves the Marcenko equation:

(2.7b)
$$g(x+y) + A(x,y) + \int_{x}^{\infty} g(y+s)A(x,s)ds = 0$$

where $x < y$.

The potential u is given by:

(2.7c)
$$u(x) = -2 \frac{d}{dx} A(x,x).$$

We note that the Marcenko equation is a *linear* integral equation. Thus, the spectral transform linearizes the kdV equation.

By solving the Marcenko equation through a generalized Cramers rule one obtains the explicit formula. first derived by Bargmann [5]:

(2.8)
$$u(x) = -2 \frac{d^2}{dx^2} \log \det(I + G(x)).$$

The determinant here is in the Fredholm sense. I and G are linear operators in the half line $[x,\infty)$. I is the identity operator and G has distribution kernel $g(r+s)$, $x \leqslant r,s < \infty$, where g is defined by (2.7a).

Of particular interest is the case of reflectionless initial data i.e. when $R(k) = 0$. In this case the operator G is of finite rank and formula (2.8) takes the form:

(2.9)
$$u_N(x) = -2 \frac{d^2}{dx^2} \log \det \left[\delta_{ij} + \frac{c_i c_j e^{-(\eta_i + \eta_j)x}}{\eta_i + \eta_j} \right].$$

where $i,j = 1,2,....,N$.

This formula, derived by Kay and Moses [10], gives the famous multisoliton solution of the kdV equation. It is the starting point of the Lax and Levermore solution of the small dispersion case.

The evolution of the spectral data with time, when $u(x,t)$ satisfies the kdV equation is particularly simple:

(2.10)
$$T(k,t) = T(k,0), \quad R(k,t) = R(k,0)e^{8ik^3 t}, \quad c_j(t) = c_j(0)e^{4\eta_j^3 t}.$$

(b)   *Initial data which are asymptotically constant as* $x \longrightarrow +\infty$:

The inverse problem in this case was solved by Buslaev and

302

Fomin [1; see also 2]. The spectrum consists of a half line with multiplicity two, as in the case of decaying potentials, plus a continuous band of multiplicity one, and can have at most finitely many discrete eigenvalues. The inverse formalism conceptually shares a lot with the formalism for decaying potentials. The spectral band of multiplicity 1 can be thought of heuristically as the continuum limit of many tightly packed discrete eigenvalues. I will not go into more detail since I will not use this formalism.

(c)   *Periodic initial data:*

The inverse problem in this case was solved by Mckean, van Moerbeke and Trubowitz [13,14,16] and by Matveev, Dubrovin and Novikov [4]. The $L_2(-\infty,\infty)$ spectrum of the operator

$$\mathcal{L} = \frac{d^2}{dx^2} - u(x) \quad \text{consists of a set of bands} \quad [\xi_1^+, \xi_0]$$

$\cup \; [\xi_2^+, \xi_1^-] \; \cup \; [\xi_3^+, \xi_2^-] \; \cup \; \ldots \quad \text{where} \quad \xi_0 > \xi_1^+ > \xi_1^- > \xi_2^+ > \xi_2^- > \ldots$ is a finite or infinite sequence of real numbers. In the finite case the last band is $(-\infty, \xi_N^-]$. We restrict ourselves to the finite case although all subsequent results generalize to the case $N = \infty$. The intervals $[\xi_i^-, \xi_i^+]$ which separate two successive spectral bands are commonly referred to as spectral gaps or simply gaps. They have decaying properties as $\xi \longrightarrow -\infty$ which depend on the smoothness of the potential u. The piece of spectral information needed in the inverse problem, in addition to the spectrum of $\mathcal{L}$ concerns the Dirichlet boundary value problem:

(2.11) $(\mathcal{L}\psi)(s) = \nu\psi(s), \quad \psi(x) = \psi(x+p) = 0,$

where x is arbitrary and p is the period of the potential. The theory of 2nd order O.D.E.'s implies that there is exactly one eigenvalue ν in each spectral gap $[\xi_i^-, \xi_i^+]$. We call this $\nu_i = \nu_i(x)$. There may be eigenvalues of (2.11) lying in the interior of the spectral bands but these are of no interest to us. (They can be correctly thought of as belonging to a spectral gap of zero width.)

The following results, obtained by the theory of 2nd order ordinary differential equations, demonstrate that sufficient data for the recovery of the potential u(x) having N spectral gaps are the values $\xi_0, \xi_1^\pm, \ldots, \xi_N^\pm$, the initial values $\nu_i(x_0)$ where $i = 1, \ldots, N$, and N initial signs (i.e. + or -) to be introduced shortly. The potential u(x) is recovered by the formula:

$$(2.12) \qquad u(x) = -\xi_0 + 2 \sum_{i=1}^{N} [\nu_i(x) - \tfrac{1}{2}(\xi_i^+ + \xi_i^-)].$$

As the arbitrary parameter x in the eigenvalue problem (2.11) varies, the evolution of the value of $\nu_i(x)$ in the gap $[\xi_i^-, \xi_i^+]$ is given by:

$$(2.13a) \qquad \frac{\partial \nu_j}{\partial x} = \frac{2 \, R(\nu_j)^{1/2}}{\displaystyle\prod_{\substack{i=1 \\ i \neq j}}^{N} (\nu_j - \nu_i)}$$

where

$$(2.13b) \qquad R(\nu) = \prod_{i=1}^{N} (\nu - \xi_i)$$

The sign of $R(\nu_j)^{1/2}$ and hence of $\dfrac{\partial \nu_j}{\partial x}$ changes when ν_j assumes the endpoint value $\xi_j^\pm$. This ensures that ν_j stays within the jth gap. We conveniently think of ν_i as taking values on two copies of the interval $[\xi_i^-, \xi_i^+]$, one for each possible sign of $\dfrac{\partial \nu_i}{\partial x}$. The endpoints of the two copies are then identified with each other and topologically ν_i varies on a circle. The position of $\nu_i(x)$ on the circle determines both its value in $[\xi^-, \xi^+]$ and the sign of $R(\nu_i)^{1/2}$.

If the potential u depends on an additional variable t in such a way that it satisfies the kdV equation, then the dependence of ν_j on t is given by:

$$(2.13c) \qquad \frac{\partial \nu_j}{\partial t} = 2(-\xi_0 - 2\nu_j + 2 \sum_{i=1}^{N} [\nu_i - \tfrac{1}{2}(\xi_i^+ + \xi_i^-)]) \frac{\partial \nu_j}{\partial x}$$

304

The system (2.13a) can be integrated exactly. To do this one divides both sides by $R(\nu_j)^{1/2}$ then multiplies by an arbitrary polynomial $Q(\nu_j)$ of degree $N - 1$ and adds over the index $j = 1,...,N$. The right hand side is shown to be independent of the ν's. The left hand side is a perfect derivative with respect to x. One integrates with respect to x. This procedure can be repeated with (2.13b) integrating with respect to t this time. One obtains:

$$(2.14a) \qquad \sum_{i=1}^{N} \int_{\nu_i(x_0,t_0)}^{\nu_i(x,t)} \frac{Q(\nu)}{R(\nu)^{1/2}} \, d\nu = a(x-x_0) + b(t-t_0).$$

If $Q(\nu) = \prod_{i=1}^{N-1} (\nu-\lambda_i)$ the coefficients a,b are:

$$(2.14b) \qquad a = 2, \quad b = -4\xi_0 - 4 \sum_{i=1}^{N} \xi_i^+ - 4 \sum_{i=1}^{N} \xi_i^- + 8 \sum_{i=1}^{N-1} \lambda_i.$$

There are N linearly independent polynomials of degree $N - 1$; therefore there are N independent equations of type (2.14). In other words (2.14) gives N "algebraic" equations for the N unknowns $\nu_i(x,t)$, provided the position of ν_i at $x = x_0$, $t = t_0$ on the ith circle is known for $i = 1,2,....,N$. In a similar manner we obtain the formula:

$$(2.15a) \qquad u(x,t) = \frac{\partial}{\partial x} \sum_{i=1}^{N} \int^{\nu_i(x,t)} \frac{P(\nu)}{R(\nu)^{1/2}} \, d\nu$$

where

$$(2.15b) \qquad P(\nu) = \nu^N - \left[\xi_0 + \sum_{i=1}^{N} \frac{\xi_i^+ + \xi_i^-}{2} \right] \nu^{N-1} +$$

$$+ \text{ arbitrary polynomial of degree N-2.}$$

One should observe that winding numbers are involved in the integrals (2.14-15). Indeed $R(\nu_i)^{1/2}$ changes sign when integration in the interval $[\xi_i^-, \xi_i^+]$ changes direction.

Solving the system (2.14) for the ν_i's is essentially the Jacobi inversion problem. The solution is obtained through the Abel transform and the theory of theta functions. The explicit solutions of (2.1) has form:

(2.16)
$$u_N(x,t) = q_N(\theta_1, \theta_2, \ldots, \theta_N; \; \xi_0, \xi_1^\pm, \ldots, \xi_N^\pm).$$

where q_N is 2π-periodic in each angle θ_i, and θ_i depends linearly on x and t i.e.

$$\theta_i(x,t) = \kappa_i x + \omega_i t - d_i.$$

The periods κ_i and the frequencies ω_i are determined uniquely by $(\xi_i^\pm - \xi_0)_{i=1}^N$. Conversely the length of all spectral bands and gaps are uniquely determined by the κ_i's and ω_i's. The d_i's are phase shifts. The beauty of this representation is that it gives the explicit solution of (2.1) in a form which decouples the N degrees of freedom in the oscillation. From a different point of view, the θ_i's are the angle variables of the kdV equation considered as a Hamiltonian system. Formula (2.16) is derived for periodic initial data. In this case the wave numbers κ_i are commensurate. The formula is still valid for noncommensurate κ_i's and gives a family of exact quasiperiodic solutions of the kdV equation.

Our method does not make use of this representation but is based on the analysis of formulas (2.14) and (2.15).

The quasiperiodic solution of (2.1) corresponding to a finite number of gaps N are referred to as quasiperiodic N-soliton solutions of the kdV equation. They are related to the decaying N-solitons discussed earlier. Indeed by letting $\xi_0 \longrightarrow \xi_1^+$, $\xi_i^- \longrightarrow \xi_{i+1}^+$ for i = 1,2,...,N-1 and $\xi_N^- = 0$ one obtains in the limit a decaying N-soliton corresponding to eigenvalues $\eta_i^2 = \xi_i^+$. Formula (2.16) reduces to the N-soliton formula (2.9). This has been shown by McKean [15].

3.   The Small Dispersion Problem

I return to the small dispersion problem (1.1). I assume that the initial data $u(x,0,\epsilon) = -v(x)$ are smooth and satisfy the normalization condition:

(3.1a)
$$\sup_{x \in \mathbb{R}} v(x) = 1, \quad \inf_{x \in \mathbb{R}} v(x) = 0.$$

Moreover $v(x)$ is either p-periodic in x, or it satisfies the decay condition

(3.1b)
$$\int_{-\infty}^{\infty} |v(x)|(|x| + 1)dx < \infty.$$

The normalization condition does not impose a real restriction. The oscillation of the solution can be regulated by the rescaling of u, x and t. The graph of the solution can be raised or lowered by the constant c through the observation that if $u(x,t)$ satisfies the kdV equation (2.1) so does $V(x,t) = c + u(x+6ct,t)$.

The case of nonpositive decaying initial data $u(x,0,\epsilon)$ was first solved by Lax and Levermore [12]. I then studied the case of nonnegative decaying $u(x,0,\epsilon)$ by an extension of the Lax-Levermore method [17]. Lax and Levermore also extend their method to step-function initial data. The more general solution is the solution of the periodic problem [19] which will be the main subject of my talk. All previous results can be derived from the solution to the periodic problem by letting the period tend to infinity. The method of solving the periodic problem has the advantage of providing more information on the solution as will be seen shortly.

In the following I assume for simplicity that the initial data have exactly one peak in each period (or in $\mathbb{R}$ in the decaying case).

307

4. The Lax–Levermore Solution: Decaying Initial Data [12]

The associated Schrödinger operator in the Lax pair is:

$$(4.1) \qquad \mathcal{L}(t) = \epsilon^2 \frac{d^2}{dx^2} - u(x,t,\epsilon)$$

where $u(x,t,\epsilon)$ is the solution of (1.1). The spectral data of the operator

$$(4.2) \qquad \mathcal{L}(0) = \epsilon^2 \frac{d^2}{dx^2} + v(x)$$

are calculated asymptotically as $\epsilon \longrightarrow 0$ by the W.K.B. method.

The spectrum of $\mathcal{L}$ at $t = 0$, which remains fixed as t increases, consists of the nonpositive half-axis of the spectral variable ξ and of a pure point spectrum, closely packed in the region $0 \leqslant \xi \leqslant 1$. I will denote the pure eigenvalues by $\eta_1^2 > \eta_2^2 > ... > \eta_N^2$. Weyl's law implies that $N = O(1/\epsilon)$. The results of the W.K.B. calculation are as follows: Let $\epsilon \longrightarrow 0$ and let the index $j(\epsilon)$ tend to infinity so that η_j tends to $\eta \in (0,1)$. Then:

$$(4.3a) \qquad \eta_j - \eta_{j+1} \sim \frac{\pi\epsilon}{\varphi(\eta)},$$

where

$$(4.3b) \qquad \varphi(\eta) = \int_{x_-(\eta)}^{x_+(\eta)} \eta(v(x)-\eta^2)^{-1/2} \, dx$$

and $x_-(\eta) < x_+(\eta)$ are the two solutions to the equation $v(x) = \eta^2$ when $0 < \eta^2 < 1$. The corresponding norming constants are given by:

$$(4.4a) \qquad \log c_j \sim \frac{1}{\epsilon} \, \theta_+(\eta), \quad j = 1,2,...,N$$

where

(4.4b) $$\theta_+(\eta) = \eta x_+(\eta) + \int_{x_+(\eta)}^{\infty} \eta - (\eta^2 - v(x))^{1/2} \, dx.$$

The evolution of the norming constants is given by formula (2.10) with t replaced by $\frac{t}{\epsilon}$. Replacing x and t by $\frac{x}{\epsilon}$ and $\frac{t}{\epsilon}$ respectively in a solution of (2.1) gives a solution of (1.1).

(4.5) $$\log c_j(t) \sim \frac{1}{\epsilon} \theta_+(\eta) + \frac{4}{\epsilon} \eta^3 t.$$

Due to the positiveness of the initial potential $u(x,0,\epsilon)$, the reflection coefficient is zero, to all W.K.B. orders.

This is an enormous simplification because it leaves us with a pure soliton solution which, however, contains $O(1/\epsilon)$ wave parameters and is extremely complicated. This is not surprising given that it represents the solution to the initial value problem (1.1) uniformly in $\mathbb{R} \times [0,T]$. It is nonoscillatory at $t = 0$ due to the very particular "phases" of the solitons at $t = 0$. As t increases, the inevitable separation of the solitons ensures the generation of oscillation.

I will show in the course of this talk that, in any rectangle on the (x,t) plane having sides $>> \epsilon$ and $<< 1$, the solution can be represented by one of the explicit quasiperiodic solutions previously discussed. The penalty for having this simple wave representation is that it is not uniform in $\mathbb{R} \times [0,T]$. As we move the rectangle on the (x,t) plane the value of the local wave parameters vary. This is modulation. Moreover, there are interfaces in the (x,t) plane across which the number of the local wave phase changes.

I proceed with the Lax Levermore solution by substituting (4.3) and (4.5) into the representation of the multisoliton solution (2.9). The determinant in (2.9) is computed as follows: The diagonal term δ_{ij} is replaced by $\lambda \delta_{ij}$, the determinant is expanded in powers of λ and then λ is set equal to 1. The determinant equals the sum:

(4.6) $$\sum_S \left\{ \prod_{i,j \in S} c_i c_j e^{-(\eta_i + \eta_j)x} \right\} \det \left[\frac{1}{\eta_i + \eta_j} \right]_{i,j \in S},$$

where S ranges over all subsets of the set of indices $\{1,2,.....N\}$. The determinant in (4.6) is a Cauchy determinant and is computed in closed form:

$$(4.7) \qquad \det\left[\frac{1}{\eta_i+\eta_j}\right]_{i,j\in S'} = \frac{\prod\limits_{i,j\in S}^{\prime} |\eta_i-\eta_j|}{\prod\limits_{i,j\in S}(\eta_i+\eta_j)}.$$

The prime on Π indicates that the factors corresponding to $i = j$ are excluded from the product. After substituting the asymptotic values of η_j and c_j calculated by the W.K.B. method one obtains:

$$(4.8) \qquad u(x,t,\epsilon) \sim -2 \frac{\partial^2}{\partial x^2} \epsilon^2 \log \Delta,$$

where

$$(4.9) \qquad \Delta = \sum_S (\tfrac{1}{2}\epsilon)^{|S|} \exp\left[-\frac{4}{\epsilon}\sum_{j\in S}(\eta_j x - 4\eta_j^3 t - \theta_+(\eta_j)) + \sum_{j\in S}\sum_{\substack{i\in S\\i\neq j}} \log \left|\frac{\eta_i-\eta_j}{\eta_i+\eta_j}\right|^2 - 2\sum_{i\in S}\log \eta_i\right]$$

Lax and Levermore now take limits in the distribution (i.e. weak) sense on both sides of (4.8) as $\epsilon \longrightarrow 0$. The limit commutes with $\frac{\partial^2}{\partial x^2}$ in the right hand side. They obtain:

$$(4.10) \qquad \bar{u}(x,t) = d - \lim_{\epsilon\to 0} u(x,t,\epsilon) = -2\frac{\partial^2}{\partial x^2}\left[d - \lim_{\epsilon\to 0}\epsilon^2 \log \Delta\right]$$

A simple estimate shows that, as $\epsilon \longrightarrow 0$, Δ is dominated by its largest term. More precisely:

$$(4.11) \qquad \epsilon^2 \log \Delta \sim 2\min_S \left\{2\epsilon\sum_{j\in S} a(\eta_j) - \epsilon^2 \sum_{j\in S}\sum_{\substack{i\in S\\i\neq j}} \log \left|\frac{\eta_i-\eta_j}{\eta_i+\eta_j}\right|\right\},$$

where

(4.12)
$$a(\eta,x,t) = \eta x - 4\eta^3 t - \theta_+(\eta).$$

Lax and Levermore replace sums by integrals over an atomic measure, by defining the distribution:

(4.13)
$$\psi(\eta;S) = \pi\epsilon \sum_{j \in S} \delta(\eta-\eta_j).$$

and writing relation (4.11) as:

(4.14)
$$\epsilon^2 \log \Delta \sim 2\min_{\delta} \{\frac{2}{\pi} \int_0^1 a(\eta)\psi(\eta)d\eta -$$
$$- \frac{1}{\pi^2} \int_0^1 \int_0^1 \log \left|\frac{\eta-\mu}{\eta+\mu}\right| \psi(\eta)\psi(\mu)d\eta d\mu\},$$

where ψ ranges over all distributions of type (4.13). As $\epsilon \longrightarrow 0$, they show that the minimization of the right hand side has the continuum limit:

(4.15)
$$d - \lim_{\epsilon \to 0} \epsilon^2 \log \Delta = Q(x,t) =$$
$$= \min_{0 \leqslant \psi \leqslant \varphi} \{\frac{4}{\pi} \int_0^1 a(\eta,x,t)\psi(\eta)d\eta - \frac{2}{\pi^2} \int_0^1 \int_0^1 \log \left|\frac{\eta-\mu}{\eta+\mu}\right| \psi(\mu)\psi(\eta)d\mu d\eta\}.$$

In the continuum limit, the distribution $\psi(\eta)$ reduces to a density which ranges over all integrable functions in $[0,1]$ and satisfies the condition $0 \leqslant \psi(\eta) \leqslant \varphi(\eta)$.

Let L denote the integral operator having distribution kernel $\frac{1}{\pi} \log \left|\frac{\eta-\mu}{\eta+\mu}\right|$ in $[0,1] \times [0,1]$. The variational equation corresponding to (4.15) is:

(4.16)
$$a(\eta,x,t) - (L\psi)(\eta) = \begin{array}{l} > 0 \Rightarrow \psi(\eta) = 0 \\ 0 \Rightarrow 0 < \psi(\eta) < \varphi(\eta) \\ < 0 \Rightarrow \psi(\eta) = \varphi(\eta) \end{array}$$

311

Lax and Levermore prove that the minimization problem has a unique solution. They obtain it by solving the variational equation. If $\psi^*(\eta,x,t)$ is its solution they easily show that:

(4.17)
$$\bar{u}(x,t) = \frac{4}{\pi} \frac{\partial}{\partial x} \int_0^1 \eta\psi^*(\eta,x,t)d\eta.$$

Before describing their solution of equation (4.16) I will derive a similar equation for the periodic problem by a different procedure.

5.   The Periodic Problem [19]

I assume $v(x)$ to be a smooth periodic function of period p which takes values in the interval $[0,1]$. The spectrum of the operator $\mathcal{L}(0) = \epsilon^2 \dfrac{d^2}{dx^2} + v(x)$ consists of a set of, in general, infinitely many spectral bands, which lie in the set $[-\infty,1]$ in the limit $\epsilon \longrightarrow 0$. The bands in the negative half axis are separated by gaps which decay exponentially as $\epsilon \longrightarrow 0$. One shows that these gaps may be ignored and that the entire negative half axis may be considered as a spectral band. This reduces the solution of the initial value problem (1.1) to a kdV solution of the finite gap type discussed earlier. An asymptotic calculation shows that the remaining bands are concentrated in the interval $[0,1]$ as $\epsilon \longrightarrow 0$ and their number $N = N(\epsilon)$ is of order $O(1/\epsilon)$. These bands have widths which are exponentially small in ϵ, while the gaps separating them have widths of order ϵ. I denote the midpoints of the bands by $n_1^2 > n_2^2 > ... > n_N^2$. As a result of the simplifying assumption of having exactly one peak in each period of $v(x)$, the asymptotic distribution of the n_j's is simple. Let the index $j = j(\epsilon)$ tend to infinity as $\epsilon \longrightarrow 0$ in such a way that $n_j \longrightarrow n$, where $0 < n < 1$. Then:

(5.1)
$$n_j - n_{j+1} \sim \frac{\pi\epsilon}{\varphi(n)}.$$

The function $\varphi(n)$ is given by the formula:

(5.2)
$$\varphi(n) = \int_{x_-(n)}^{x_+(n)} n(v(x)-n^2)^{-1/2}\, dx \geq 0.$$

where $x_+(n)$ are defined by the figure 5.1:

(5.3)

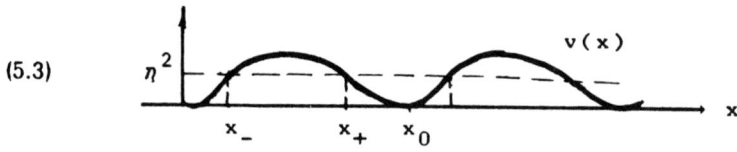

Figure 5.1

313

The band and gap structure of the spectrum is given in terms of the midpoints of the bands n_j^2 and their widths $2\delta_j$:

(5.4a) $$\varepsilon_j^+ = n_j^2 - \delta_j, \quad \varepsilon_j^- = n_{j+1}^2 + \delta_{j+1}.$$

(5.4b) We can assume that $\varepsilon_N^- = 0$.

The width $2\delta_j$ of the jth band is given asymptotically as $\varepsilon \longrightarrow 0$ by:

(5.5) $$\gamma_j = -\varepsilon \log \delta_j \sim \int_{x_+(\eta)}^{x_-(\eta)+p} (\eta^2 - v(x))^{1/2} dx = \gamma(\eta) \geq 0.$$

The asymptotic results quoted are derived by the W.K.B. method.

The solution of the initial value problem (1.1) will be derived from equations (2.14–15). One should recall that equation (2.15) gives $u(x,t)$ in terms of the N numbers $\nu_i(x,t)$. For the polynomial $P(\nu)$ in this equation I make the choice:

(5.6) $$P(\nu) = \prod_{i=1}^{N} (\nu - n_i^2).$$

The N unknowns $\nu_i(x,t)$ satisfy N equation obtained from (2.14) for N choices of the polynomial $Q(\nu)$. These choices are:

(5.7) $$P_j(\nu) = \frac{P(\nu)}{\nu - n_j^2} \text{ where } j = 1,2,...,N.$$

The coefficients $a = a_j$ and $b = b_j$ in (2.14b) corresponding to the choice $Q(\nu) = P_j(\nu)$ are given by:

(5.8) $$a_j = 2n_j, \quad b_j = -8n_j^3$$

The advantage of my choice for the polynomials $P(\nu)$ and $Q(\nu)$ lies into the enormous amount of cancelation in the quotient $P(\nu)/R(\nu)^{1/2}$. This allows the integrations in (2.14–15) to be performed in closed form. Indeed ignoring signs:

(5.9)
$$\left|\frac{P(\nu)}{R(\nu)^{1/2}}\right| = (\nu - \varepsilon_N^-)^{-1/2} \prod_{i=1}^{N} \frac{|\nu-\eta_i^2|}{|\nu-\varepsilon_i^+|^{1/2}|\nu-\varepsilon_i^-|^{1/2}} =$$

$$= \nu^{-1/2} \prod_{i=1}^{N} \frac{|\nu-\eta_i^2|}{(\nu-\eta_i^2)^2-\delta_i^2}.$$

Due to the exponentially small size of δ_i the product in the right is essentially equal to 1 except for values of ν which make $|\nu-\eta_i^2|$ comparable to δ_i^2 for some i. Even in this case, only the ith factor in (5.9) is significant.

The calculation of the integrals in (2.14-15) is further simplified by the topological properties of the graphs of the $\nu_i(x,t)$ illustrated by figure 5.2 at t = 0. In this figure the horizontal lines represent the spectral bands. The dark full line represents the graph of a typical $\nu_i(x,t)$ for fixed t.

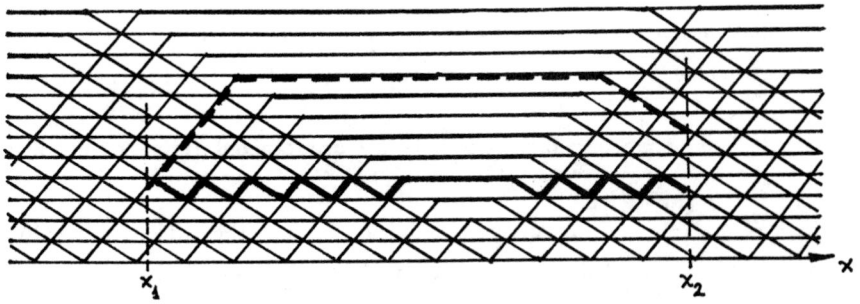

Figure 5.2

The following "topological" properties can be proven by a careful W.K.B. calculation:

(i)   In the region of (x,ν_i) where $v(x) > \nu_i$ the graph of ν_i is an oscillation between the ith and $i + 1^{st}$ band. The period of the oscillation has order $O(\epsilon)$. The points in the (x,ν) plane at which the graphs of $\nu_i(x)$ and $\nu_{i-1}(x)$ touch the ith band appear in pairs and the distance between them decays exponentially as $\epsilon \longrightarrow 0$.

Neglecting the exponentially small error we can think of the two graphs as intersecting each other on the ith band; hence the term topological property.

(ii)   In the x–interval in which $v(x) > \nu_i$ the value of ν_i performs exactly one cycle. It assumes its maximal value at exactly one x. In the whole x–interval, apart from pieces of order ϵ close to the endpoints, its distance from its maximal value is exponentially small in ϵ.

A simple study of the evolution equations (2.13c) for ν_i, $i = 1,...,N$, shows that the topology is preserved for $t > 0$.

The left hand side of equations (2.14–15) involves the sum of integration over the graphs of all $\nu_i(x,t)$ (see figure 5.2). The topology of the graphs makes this equal to the sum of integrations over all paths of the form of the dark broken line in figure 5.2. The advantage of the latter is that they have small winding numbers, as opposed to winding numbers of order $1/\epsilon$ of a graph of ν. The calculation is simplified if one still integrates over the graphs of the ν_i's but assigns to them *effective winding numbers*. Thus, the effective winding number of the graph of ν_i is the number of times $\dfrac{\partial \nu_i}{\partial x}$ changes sign without a simultaneous change of sign $\dfrac{\partial \nu_{i+1}}{\partial x}$ or $\dfrac{\partial \nu_{i-1}}{\partial x}$. In diagram 5.2 for example the effective winding number of the solid dark path when $x \in [x_1,x_2]$ equals 1.

Having given a flavor of the detail I proceed to write the result of the integration in equations (2.14–15). Equation (2.15a) can be written

$$(5.10) \qquad \int^x u(x',t,\epsilon)dx' \sim -2 \sum_{i=1}^{N(\epsilon)} \epsilon \eta_i \sigma_i(x,t,\epsilon) + o(1) \text{ as } \epsilon \longrightarrow 0.$$

Here $\sigma_i(x,t,\epsilon)$ is defined by

$$(5.11) \qquad \sigma_i(x,t,\epsilon) = - \text{ sgn } \frac{P(\nu_i(x,t,\epsilon))}{R(\nu_i(x,t,\epsilon))^{1/2}} = \text{ sgn } \frac{\partial \nu_i}{\partial x}.$$

The system of equations (2.14) becomes after the asymptotic evaluation of the integrals:

$$\left\{\sum_{\substack{i=1 \\ i \neq j \\ i \neq j-1}}^{N(\epsilon)} \epsilon \left[\frac{1}{2}(1 - \sigma_i) - w_i\right] \log \left|\frac{\eta_i - \eta_j}{\eta_i + \eta_j}\right| + w_j \Upsilon(\eta_j) + r_j\right\}\Bigg|_{(x_0,t_0)}^{(x,t)} =$$

(5.12)

$$= \eta(x-x_0) - 4\eta^3(t-t_0) + o(1) \text{ as } \epsilon \longrightarrow 0.$$

The function $\Upsilon(\eta)$ is defined by relation (5.5). $w_i(x,t,\epsilon) - w_i(x_0,t_0,\epsilon)$ is the effective winding number of ν_i between points (x_0,t_0) and (x,t). The integer valued function $w_i(x_0,t_0)$ can be chosen arbitrarily. The integrals corresponding to $i = j$ and $i = j - 1$ have been excluded from the sum in (5.12). Their contribution is given by the function r_j:

$$r_j(x,t,\epsilon) = -\frac{\epsilon\sigma_j}{2} \log \{|\nu_j - \eta_j^2| + [(\nu_j - \eta_j^2)^2 - \delta_j^2]^{1/2}\} -$$

(5.13)

$$\frac{\epsilon\sigma_{j-1}}{2} \log \{|\nu_{j-1} - \eta_j^2| + [(\nu_j - \eta_j^2)^2 - \delta_j^2]^{1/2}\}.$$

One easily sees that each of the terms on the right hand side of (5.13) is bounded by $\Upsilon(\eta_j)$. The sharper estimate

(5.14) $$|r_j(x,t,\epsilon)| \leq \frac{1}{2}\Upsilon(\eta_j) + o(1) \text{ as } \epsilon \longrightarrow 0$$

is obtained after a short calculation which makes use of the topology of the graphs of $\nu_i(x,t,\epsilon)$. It is easy to see that $r_j(x,t,\epsilon)$ is negligible unless either $\log|\nu_j - \eta_j^2|$ or $\log|\nu_{j-1} - \eta_j^2|$ are of order $\frac{1}{\epsilon}$. In this case either ν_j or ν_{j-1} takes a value exponentially close to η_j^2.

Following the spirit of the Lax–Levermore solution I introduce the distribution:

(5.15) $\psi_\epsilon(\eta,x,t) = \pi\epsilon \sum_{i=1}^{N} \left[\frac{1}{2}(1 - \sigma_i(x,t,\epsilon)) - w_i(x,t,\epsilon)\right] \delta(\eta-\eta_i).$

Equation (5.12) is written

$$\left\{\frac{1}{\pi}\int_0^1{}^* \log\left|\frac{\eta_j-\mu}{\eta_j+\mu}\right| \psi_\epsilon(\mu)d\mu + w_j\Upsilon(\eta_j) + r_j\right\}\Bigg|_{(x_0,t_0)}^{(x,t)} =$$

(5.16)

$$= \eta_j(x-x_0) + 4\eta_j^3(t-t_0) + o(1) \text{ as } \epsilon \longrightarrow 0.$$

The asterisk in the integral represents the exclusion of the terms corresponding to $i = j$ and $i = j - 1$ from the sum in (5.12). Still in the spirit of the Lax-Levermore argument one shows:

(5.17) $\qquad\qquad \psi_\epsilon(\eta,x,t) \longrightarrow \psi(\eta,x,t) \text{ as } \epsilon \longrightarrow 0,$

in the distribution sense, where ψ is a measurable bounded function. Let now $\epsilon \longrightarrow 0$ and $j(\epsilon) \longrightarrow \infty$ in such a way that $\eta_j \longrightarrow \eta$. Equation (5.16) implies that the limit $w_j\Upsilon(\eta_j) + r_j$ exists. It is easily shown to be a continuous function of η as a consequence of the boundedness of $\psi(\eta,x,t)$. We set:

(5.18a) $\qquad\qquad \lim_{\epsilon\to 0}(w_j\Upsilon(\eta_j) + r_j) = w(\eta,x,t)\Upsilon(\eta) + r(\eta,x,t),$

in which the integer valued function $w(\eta,x,t)$ and the remainder $r(\eta,x,t)$ are uniquely determined by the requirement:

(5.18b) $\qquad\qquad |r(\eta,x,t)| \leqslant \frac{1}{2}\Upsilon(\eta).$

We clearly have:

(5.19) $\qquad\qquad w_j \longrightarrow w(\eta,x,t) \qquad r_j \longrightarrow r(\eta,x,t)$

as $\epsilon \longrightarrow 0$, almost everywhere in x,t.

The limit of equations (5.16) as $\epsilon \longrightarrow 0$ is:

318

(5.20a) $\qquad (L\psi)(\eta,x,t) + w(\eta,x,t)\Upsilon(\eta) + r(\eta,x,t) = a(\eta,x,t)$

where

(5.20b) $\quad a(\eta,x,t) = \eta(x-x_0) - 4\eta^3(t-t_0) + (L\psi)(\eta,x_0,t_0) + w(\eta,x_0,t_0)\Upsilon(\eta) + r(\eta,x_0,t_0).$

I select $t_0 = 0$ and x_0 as in diagram 5.1.

A W.K.B. calculation gives:

$$\psi(\eta,x_0,0) = 0$$

(5.20c)

$$w(\eta,x_0,0)\Upsilon(\eta) + r(\eta,x_0,0) = \int_{x_+(\eta)}^{x_0} (\eta^2-v(x))^{1/2}dx$$

Substituting this into equation (5.20b) one obtains an explicit expression for the function $a(\eta,x,t)$ in terms of the initial data $v(x)$:

(5.20d) $\qquad a(\eta,x,t) = \eta x - 4\eta^3 t - \eta x_0 + \int_{x_+(\eta)}^{x_0} (\eta^2 - v(x))^{1/2}dx$.

The unknown function $r(\eta,x,t)$ can be eliminated from equation (5.20a) due to the following crucial observation when the limit $\epsilon \longrightarrow 0$ is taken in (5.13): When $r = r(\eta_j,x,t) \neq 0$ the relations:

(5.21) $\qquad \lim_{\epsilon\to 0} \dfrac{|\nu_i - \eta_j^2|}{e^{-r/\epsilon}} \leq 1$ and $\lim_{\epsilon\to 0}\sigma_i = \text{sgn } r,$

hold simultaneously either for $i = j$ or for $i = j - 1$. The definition of ψ and the second relation (5.21) imply:

(5.22a) $\qquad r(\eta) < 0 \Rightarrow \psi(\eta) = (1 - w(\eta))\varphi(\eta)$

(5.22b) $\qquad r(\eta) > 0 \Rightarrow \psi(\eta) = -w(\eta)\varphi(\eta)$

319

From these and (5.20a) one infers easily:

$$-\tfrac{1}{2}\Upsilon < a - L\psi - w\Upsilon < 0 \Rightarrow \psi = (1 - w)\varphi$$

(5.23)
$$a - L\psi - w\Upsilon = 0 \Rightarrow -w\varphi < \psi < (1 - w)\varphi$$

$$0 < a - L\psi - w\Upsilon < \tfrac{1}{2}\Upsilon \Rightarrow \psi = -w\varphi$$

Relation (5.23) can be simplified to:

(5.24)
$$\Upsilon\Upsilon < a - L\psi < (1 + \Upsilon)\Upsilon \Rightarrow \psi = -\Upsilon\varphi$$

$$a - L\psi = \Upsilon\Upsilon \Rightarrow -\Upsilon\varphi < \psi < (1 - \Upsilon)\varphi$$

where Υ is an integer. A comparison between relations (5.23) and (5.24) gives:

(5.25)
$$w = \Upsilon + 1 \quad \text{when} \quad (\Upsilon + \tfrac{1}{2})\Upsilon < a - L\psi \leq (\Upsilon + 1)\Upsilon$$

$$w = \Upsilon \qquad \text{when} \quad \Upsilon\Upsilon \leq a - L\psi < (\Upsilon + \tfrac{1}{2})\Upsilon.$$

I call relation (5.24) the *basic integral equation* of the continuum limit. As will be seen shortly it determines the unknown function $\psi(\eta,x,t)$ uniquely. The function $w(\eta,x,t)$ is then determined from (5.25). Equation (5.24) is the analogue of the Lax-Levermore variational equation (4.16). In fact the latter follows from the former if we let the period $p \longrightarrow \infty$ in an appropriate way. This amounts to setting $w = 0$ and then letting $\Upsilon(\eta) \longrightarrow \infty$ in (5.23).

The weak limit of the solution $u(x,t,\epsilon)$ of (1.1) with periodic initial data is obtained by taking the limit $\epsilon \longrightarrow 0$ in (5.10).

(5.26) $$\lim_{\epsilon \to 0} \int^{x} u(x',t,\epsilon)dx' = \frac{4}{\pi} \int_{0}^{1} \eta \{\psi(\eta,x,t) + w(\eta,x,t)\varphi(\eta)\}d\eta.$$

This implies

(5.27) $$\bar{u}(x,t) = d - \lim_{\epsilon \to 0} u(x,t,\epsilon) = \frac{4}{\pi}\frac{\partial}{\partial x} \int_{0}^{1} \eta\{\psi(\eta,x,t) + w(\eta,x,t)\varphi(\eta)\}d\eta.$$

320

6. The Solution of the Basic Integral Equation

I will now give a fairly detailed solution to the basic integral equation (5.24) following the method developed by Lax and Levermore. I first extend $\psi(\eta,x,t)$ so that it is defined for each real value of η:

(6.1)
$$\psi(\eta,x,t) = 0 \qquad \text{when } |\eta| > 1$$
$$\psi(\eta,x,t) = -\psi(-\eta,x,t)$$

The operator L is written:

(6.2) $\quad (L\psi)(\eta) = \dfrac{1}{\pi} \displaystyle\int_0^1 \log\left|\dfrac{\eta-\mu}{\eta+\mu}\right| \psi(\mu)d\mu = \dfrac{1}{\pi} \displaystyle\int_{-1}^1 \log|\eta-\mu|\, \psi(\mu)d\mu.$

I will express the basic integral equation (5.24) in terms of the function:

(6.3) $\qquad F(\varsigma,x,t) = i\varsigma(x-x_0) - 4i\varsigma^3 t - \dfrac{i}{\pi} \displaystyle\int_{-1}^1 \log(\varsigma-\mu)\psi(\mu,x,t)d\mu$

which is holomorphic in the variable ς in the complex upper half-plane. F and its partial derivatives have the following decay as $\varsigma \longrightarrow \infty$.

(6.4) $\qquad F(\varsigma,x,t) = i\varsigma(x-x_0) - 4i\varsigma^3 t + O\left(\dfrac{1}{|\varsigma|}\right)$

(6.5) $\qquad F_x(\varsigma,x,t) = i\varsigma + O\left(\dfrac{1}{|\varsigma|}\right), \quad F_t(\varsigma,x,t) = -4i\varsigma^3 + O\left(\dfrac{1}{|\varsigma|}\right)$

(6.6) $\quad F_{x\varsigma}(\varsigma,x,t) = i + O\left(\dfrac{1}{|\varsigma|^2}\right), \quad F_{t\varsigma}(\varsigma,x,t) = -12i\varsigma^2 + O\left(\dfrac{1}{|\varsigma|^2}\right)$

The boundary values of these functions on the real axis are easily computed:

(6.7) $\quad F(\eta+i0,x,t) = \int_{\eta}^{1} \psi(\mu,x,t)d\mu + i[\eta(x-x_0)-4\eta^3 t-(L\psi)(\eta,x,t)]$

(6.8a) $\quad F_x(\eta+i0,x,t) = \int_{\eta}^{1} \psi_x(\mu,x,t)d\mu + i[\eta-(L\psi_x)(\eta,x,t)]$

(6.8b) $\quad F_t(\eta+i0,x,t) = \int_{\eta}^{1} \psi_t(\mu,x,t)d\mu - i[4\eta^3+(L\psi_t)(\eta,x,t)]$

(6.8c) $\quad F_{x\zeta}(\eta+i0,x,t) = -\psi_x(\eta,x,t)+i[1-(H\psi_x)(\eta,x,t)]$

(6.8d) $\quad F_{t\zeta}(\eta+i0,x,t) = -\psi_t(\eta,x,t)-i[12\eta^2+(H\psi_t)(\eta,x,t)]$

where H is the Hilbert transform. Knowledge of $F(\zeta,x,t)$ implies knowledge of $\psi(\eta,x,t)$. Indeed, it follows from (6.7) that ψ can be expressed in terms of F:

(6.9) $\qquad\qquad \psi(\eta,x,t) = -\text{Re}\, \dfrac{\partial F}{\partial \eta}\,(\eta + i0,x,t).$

Furthermore, the function $a - L\psi$ which appears in the basic integral equation (5.24) (a is defined by (5.20d)) can be expressed in terms of F by the use of (6.7):

(6.10a) $\qquad\qquad \begin{aligned} a(\eta,x,t) \;-\; L\psi(\eta,x,t) \;&= \\ &= \text{Im}\; F(\eta+i0,x,t) \;+\; b(\eta), \end{aligned}$

where the function:

(6.10b) $\qquad\qquad b(\eta) = \int_{x_+(\eta)}^{x_0} (\eta^2-v(x))^{1/2}dx \geq 0$

is determined from the initial data. The points x_0 and $x_+(\eta)$ are defined in figure 5.1.

The basic integral equation (5.24) is rewritten in terms of F as:

322

$$\text{(6.11a)} \qquad \begin{aligned} \tau\Upsilon(\eta) < \text{Im } F(\eta+i0,x,t) + b(\eta) &< (1+\tau)\Upsilon(\eta) \Rightarrow \\ \Rightarrow \text{Re } \frac{\partial F}{\partial \eta}(\eta+i0,x,t) &= \tau\varphi(\eta) \end{aligned}$$

$$\text{(6.11b)} \qquad \begin{aligned} \text{Im } F(\eta+i0,x,t) + b(\eta) &= \tau\Upsilon(\eta) \\ \Rightarrow (\tau-1)\varphi(\eta) < \text{Re } \frac{\partial F}{\partial \eta}(\eta+i0,x,t) &< \tau\varphi(\eta). \end{aligned}$$

The functions $\varphi(\eta)$, $\Upsilon(\eta)$ and $b(\eta)$ are given by equations (5.2), (5.5) and (6.10b) respectively.

At $t = 0$ the solution to equations (6.11a-b) is obtained by inspection:

$$\text{(6.12)} \qquad F(\zeta,x,0) = - \int_{x_0}^{x} (v(x')-\zeta^2)^{1/2} dx',$$

where the branch cut for the square root is taken to be the real negative half-axis. Thus at $t = 0$, the graph of the function $\eta^2 = v(x)$ can be characterized as the common boundary of two regions in the (x,η^2) plane namely the region $\eta^2 < v(x)$ in which the assumption of relation (6.11a) is true while the assumption of relation (6.11b) is false and the region $v(x) < \eta^2 < 1$ in which the reverse is true.

Lax and Levermore obtain the solution of (6.11a-b) for $t > 0$ by determining the time evolution of this boundary curve. Following their method, I assume, with some forethought, that there are (x,t) regions in which this curve is the graph of a multivalued function of x. I denote these multiple values by

$$\beta_0^2(x,t) > \beta_1^2(x,t) > \dots > \beta_{2m}^2(x,t) \geqslant 0,$$

where the integer $m = m(x,t)$ depends on x and t. (My $\beta_0, \beta_1, \dots, \beta_{2m}$ are the $\beta_1, \dots, \beta_n$ of Lax and Levermore.) The assumption of (6.11b) holds in the set:

$$\text{(6.13)} \qquad I(x,t): \{\eta: (\eta^2-\beta_0^2)(\eta^2-\beta_1^2)\dots(\eta^2-\beta_{2m}^2) < 0\}$$

while the assumption of (6.11a) holds outside I(x,t).

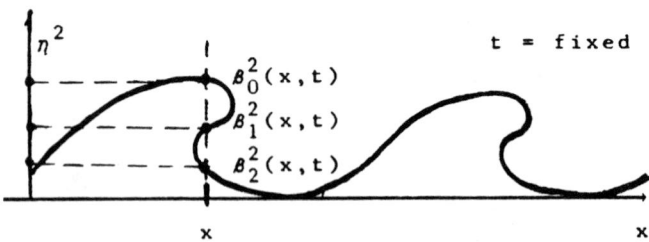

Fig. 6.1

Differentiating (6.11a-b) with respect to x and t and substituting in (6.8a) and (6.8b) I obtain:

(6.14a) $\text{Im } F_x(\eta+i0,x,t) = 0$, $\text{Im } F_t(\eta+i0,x,t) = 0$ when $\eta \in I(x,t)$

(6.14b) $\dfrac{d}{d\eta} \text{ Re } F_x(\eta+i0,x,t) = 0$, $\dfrac{d}{d\eta} \text{ Re } F_t(\eta+i0,x,t) = 0$ when $\eta \notin I(x,t)$.

Given the set I, there cannot be two holomorphic functions for F_x which satisfy (6.14), (6.5) and (6.6). If there were, their difference G would satisfy:

(6.15)
$$\text{Im } G_\zeta(\eta+i0) = 0 \text{ on } I, \quad \text{Re } G_\zeta(\eta+i0) = 0 \text{ off } I.$$
$$G_\zeta = O\left[\frac{1}{|\zeta|^2}\right] \text{ as } \zeta \longrightarrow \infty$$

The function G_ζ^2 would be holomorphic in the upper half plane, have zero imaginary part on the real axis and have order $O\left[\dfrac{1}{|\zeta|^4}\right]$ as $\zeta \longrightarrow \infty$. Thus G_ζ would be zero. This would imply $G = 0$. Similarly there cannot be two functions F_t which satisfy (6.14), (6.5) and (6.6). Thus, if one can produce functions which satisfy (6.14), (6.5), (6.6), they will be the only possible candidates for F_x and F_t. To produce such functions Lax and Levermore utilize the expression:

(6.16)
$$\mathcal{R}^{1/2}(\zeta) = \left\{ \prod_{k=0}^{2m} (\beta_k^2 - \zeta^2) \right\}^{1/2}$$

which has zero imaginary part on I and zero real part off I. They

324

choose the sign so that $i\mathfrak{R}^{1/2}(\mathfrak{s}) > 0$ when $\mathfrak{s}$ is large and positive. Eventually $\mathfrak{R}$ will depend on x and t through the dependence of the β_k's on x and t. (Lax and Levermore call $\mathfrak{R}$ what I call $\mathfrak{R}^{1/2}$).

It is easily verified that the expressions:

$$(6.17) \qquad F_x(\mathfrak{s},x,t) = \int_0^{\mathfrak{s}} \frac{\mathcal{P}(\mathfrak{s}')}{\mathfrak{R}(\mathfrak{s}')^{1/2}} \, d\mathfrak{s}',$$

$$(6.18) \qquad F_t(\mathfrak{s},x,t) = \int_0^{\mathfrak{s}} \frac{\mathfrak{Q}(\mathfrak{s}')}{\mathfrak{R}(\mathfrak{s}')^{1/2}} \, d\mathfrak{s}',$$

satisfy relations (6.14), (6.5) and (6.6) when $\mathcal{P}(\mathfrak{s})$ and $\mathfrak{Q}(\mathfrak{s})$ are odd polynomials given by:

$$(6.19) \qquad \mathcal{P}(\mathfrak{s}) = \mathfrak{s} \prod_{i=1}^{m} (\mathfrak{s}^2 - \sigma_i^2), \quad \mathfrak{Q}(\mathfrak{s}) = -12\mathfrak{s} \prod_{i=1}^{m+1} (\mathfrak{s}^2 - \rho_i^2),$$

and the m real numbers σ_i are determined uniquely by the m equations:

$$(6.20) \qquad \int_{\beta_{2k}}^{\beta_{2k-1}} \frac{\mathcal{P}(\mathfrak{s})}{\mathfrak{R}(\mathfrak{s})^{1/2}} \, d\mathfrak{s} = 0 \qquad k = 1,2,\ldots,m,$$

while the m+1 real numbers ρ_i are determined uniquely by the m+1 equations:

$$(6.21) \qquad \int_{\beta_{2k}}^{\beta_{2k-1}} \frac{\mathfrak{Q}(\mathfrak{s})}{\mathfrak{R}(\mathfrak{s})^{1/2}} \, d\mathfrak{s} = 0 \qquad k = 1,2,\ldots,m \text{ and}$$

$$(6.22) \qquad \sum_{i=1}^{m+1} \rho_i^2 = \frac{1}{2} \sum_{j=0}^{2m} \beta_j^2.$$

The idea which leads to these expressions is simple. $\frac{\mathcal{P}(\mathfrak{s})}{\mathfrak{R}(\mathfrak{s})^{1/2}}$ is pure real on I and pure imaginary off I. When we integrate it from

$0 + i0$ to $\eta + i0$ to obtain F_x, it still remains pure real on I due to relations (6.20). Thus (6.14) is satisfied. The degree of $\mathcal{P}$ has been chosen so that F_x satisfies (6.5) and (6.6). The justification of the expression for F_t is similar.

The polynomials $\mathcal{P}$, $\mathcal{Q}$ and $\mathcal{R}$ are only known up to the number and values of the parameters $\beta_1^2, \beta_2^2, \ldots, \beta_{2m}^2$. These parameters are obtained from the consistency condition $F_{xt} = F_{tx}$ which is equivalent to:

(6.23a)
$$\frac{\partial}{\partial t} \frac{\mathcal{P}(\varsigma)}{\mathcal{R}(\varsigma)^{1/2}} + \frac{\partial}{\partial x} \frac{\mathcal{Q}(\varsigma)}{\mathcal{R}(\varsigma)^{1/2}} = 0.$$

This implies:

(6.23b)
$$(\mathcal{P}_t + \mathcal{Q}_x)\mathcal{R} - \frac{1}{2}(\mathcal{P}\mathcal{R}_t + \mathcal{Q}\mathcal{R}_x) = 0.$$

When (6.23b) is evaluated at $\varsigma = \beta_j$ the first term drops out since $\mathcal{R}(\beta_j) = 0$. The second term is manipulated easily to give the following evolution equations for the β_j's.

(6.24)
$$\frac{\partial}{\partial t}\beta_j^2 + \frac{\mathcal{Q}(\beta_j; \beta^2)}{\mathcal{P}(\beta_j; \beta^2)}\frac{\partial}{\partial x}\beta_j^2 = 0, \quad j = 1, \ldots, m.$$

Here, $\mathcal{P}(\beta_j; \beta^2)$ and $\mathcal{Q}(\beta_j; \beta^2)$ are the polynomials $\mathcal{P}(\varsigma)$ and $\mathcal{Q}(\varsigma)$ evaluated at $\varsigma = \beta_j$; the second variable in this notation serves as a reminder of the fact that $\mathcal{P}$ and $\mathcal{Q}$ depend on all the β_k^2's, which I denote by $\beta^2 = \beta^2(x,t)$ collectively. The denominator $\mathcal{P}(\beta_j; \beta^2)$ is non-zero because the constraints (6.20) imply that there is exactly one root σ_i^2 in each open interval $(\beta_{2k}^2, \beta_{2k-1}^2)$.

It is not difficult to prove that conditions (6.24) are also sufficient for the validity of (6.23a) and hence for the compatibility of the expressions (6.17-18) for F_x and F_t.

Lax and Levermore write the equations (6.24) for the evolution of the β_k^2's in characteristic form:

(6.25a)
$$\frac{d\eta}{dt} = 0, \quad \frac{dx}{dt} = \frac{2(\eta; \beta^2(x,t))}{\mathcal{P}(\eta; \beta^2(x,t))}$$

with the initial condition

(6.25b)
$$\eta^2(0) = v(x(0)),$$

where $x(0)$ ranges over the set of real numbers. Equations (6.25a-b) give the time evolution of the curve in the (x, η^2) plane which is the common boundary of the regions in which the assumptions of relation (6.11a) or (6.11b) hold. We have seen that at $t = 0$ this curve is the graph of the function $\eta^2 = v(x)$. Thus $\beta_0^2(x,0) = v(x)$ and there are no $\beta_1^2, \beta_2^2 \ldots$ at $t = 0$.

In this case (6.19) and (6.22) imply:

(6.26)
$$\mathcal{P}(\mathfrak{z}) = \mathfrak{z}, \quad \mathfrak{Q}(\mathfrak{z}) = -12\mathfrak{z}^3 + 6\beta_0^2 \mathfrak{z}.$$

Equations (6.25a) become:

(6.27a)
$$\frac{d\eta}{dt} = 0, \quad \frac{dx}{dt} = -12\eta^2 + 6\beta_0^2 = -6\eta^2,$$

the last equality following from

(6.27b)
$$\beta_0^2(x(t),t) = \eta^2(t),$$

which is valid as long as there are no β_1^2, β_2^2 etc. Points of the curve further from the x axis travel at higher speeds. Therefore, after a finite amount of time which is called *breaktime*, a region in the x variable appears in which the curve is the graph of a multivalued function of x. In this region there are more β_k^2's than one and the simple expressions (6.26) for $\mathcal{P}$ and $\mathfrak{Q}$ are no longer valid. The correct expressions obtained from (6.19) involve $\beta_0^2, \beta_1^2, \beta_2^2$ etc. The expression for the speed $\frac{dx}{dt}$ becomes non-local because it involves all the points of the curve which have the same abscissa. The evolution equation can no longer be solved explicitly.

The knowledge of the curve and hence of the β_k^2's implies, through (6.17) and (6.18) the knowledge of the functions $F_x(\zeta,x,t)$ and $F_t(\zeta,x,t)$. Since $F(\zeta,x,0)$ is known $F(\zeta,x,t)$ follows uniquely:

$$
\begin{aligned}
F(\zeta,x,t) &= F(\zeta,x,0) + \int_0^t F_t(\zeta,x,t')dt' = \\
&= -\int_{x_0}^{x} (v(x')-\zeta^2)^{1/2}dx' + \\
&\quad + \int_0^t\int_0^\zeta \frac{\mathfrak{Q}(\zeta';\beta(x,t'))}{\mathfrak{R}(\zeta';\beta(x,t'))} d\zeta'dt'
\end{aligned}
$$

(6.28)

The function $\psi(\eta,x,t)$ is obtained from $F(\zeta,x,t)$ by (6.7).

328

7. Capturing the Wave Form

I consider the case when ϵ is small but non-zero. I recall that the solution of (1.1) is given by the formula:

$$(7.1) \qquad u(x,t,\epsilon) = -\xi_0 + 2 \sum_{i=1}^{\infty} (\nu_i(x,t,\epsilon) - \tfrac{1}{2}(\xi_i^+ + \xi_i^-)).$$

Recalling that the terms corresponding to $\nu_i < 0$ are negligible and that the widths of the bands decay exponentially as $\epsilon \longrightarrow 0$ I write:

$$(7.2) \qquad u(x,t,\epsilon) \sim -1 + 2 \sum_{i=1}^{N(\epsilon)} (\nu_i(x,t,\epsilon) - \tfrac{1}{2}(\eta_i^2 + \eta_{i+1}^2)).$$

The functions $\nu_i(x,t,\epsilon)$ behave differently in the two regions defined by $r(\nu_i,x,t) \neq 0$ and $r(\nu_i,x,t) = 0$, where $r(\cdot,x,t)$ is the remainder function introduced by (5.18). According to our analysis of the continuum limit these two regions are given in terms of the β_i's as follows:

$$(7.3) \qquad \begin{array}{l} r(\eta,x,t) \neq 0 \text{ when } \beta_{2k}^2 < \eta^2 < \beta_{2k-1}^2 \qquad k = 0,1,2,\ldots,m \\ \qquad\qquad (\text{I have set } \beta_{-1}^2 = 1), \\ r(\eta,x,t) = 0 \text{ otherwise.} \end{array}$$

In the region $r \neq 0$ relation (5.21) implies that each η_j^2 has exponentially small distance either from ν_j or from ν_{j-1}. It follows (see diagram 7.1) that ν_i is within an exponentially small distance from η_i^2 when $\beta_0^2 < \eta_i^2 < 1$. On the other hand, in the interval $(\beta_{2k}^2, \beta_{2k-1}^2)$ for $k = 1,\ldots,m$ there can be at most one value ν_i which is not constrained to be at an exponentially small distance from either η_i^2 or η_{i+1}^2. This is shown in figure 7.1. We denote this value by $\mathcal{n}_k$.

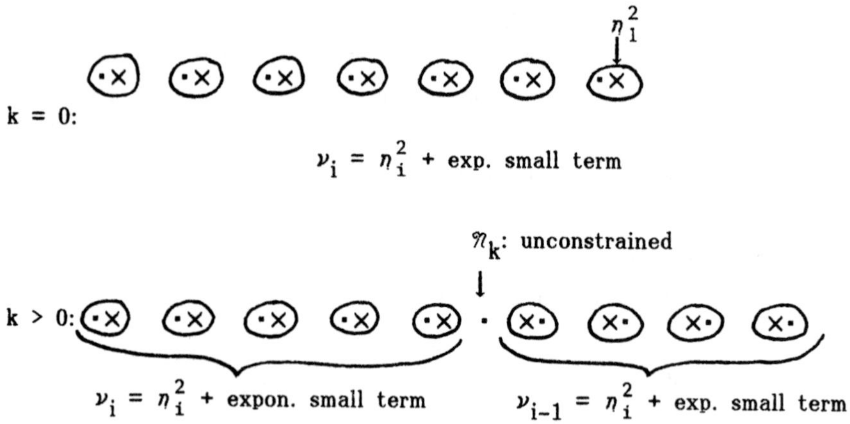

Fig. 7.1

In this diagram the crosses represent the n_i^2's and the dots represent the ν_i's. A cross and a dot in a circle indicates that the difference of the corresponding values is exponentially small.

The contribution to the series in (7.2) of the terms having $\nu_i \in (\beta_{2k-1}^2, \beta_{2k}^2)$ is given up to an exponentially small error by:

(7.4)
$$1 - \beta_0^2 \qquad \text{when } k=0$$
$$-(\beta_{2k-1}^2 - \mathcal{n}_k) + \mathcal{n}_k - \beta_{2k}^2 = 2[\mathcal{n}_k - \frac{1}{2}(\beta_{2k}^2 + \beta_{2k-1}^2)] \qquad \text{when } k>0$$

I now examine the behavior of ν_i in the region $r(\nu_i, x, t) = 0$. Before breaktime, this region is confined to the interval $(0, \beta_0^2)$; there are no other β's. A W.K.B. calculation shows that before breaktime the following *averaging principle* holds: Let $\ell = \ell(\epsilon) > 0$ satisfy: $\epsilon \ll \ell(\epsilon) \ll 1$ as $\epsilon \longrightarrow 0$. Then for all n satisfying $r(n, x, t) = 0$:

(7.5)
$$\sum_{\{i: \, |n_i^2 - n^2| < \ell(\epsilon)\}} [\nu_i(x, t, \epsilon) - \frac{1}{2}(n_i^2 + n_{i+1}^2)] \ll \ell(\epsilon) \text{ as } \epsilon \longrightarrow 0.$$

I make the assumption that this averaging principle holds also beyond breaktime. It follows directly that the

330

contribution from the terms of (7.2) having ν_i's which satisfy $r(\nu_i,x,t) = 0$ or equivalently

$\nu_i \in (0,B_{2m}^2) \cup (B_{2m-1}^2, B_{2m-2}^2) \cup \ldots \cup (B_1^2, B_0^2)$ is zero.

I substitute (7.4) into (7.2) to obtain:

(7.6) $\quad u(x,t,\epsilon) \sim -B_0^2 + 2 \sum_{k=1}^{m(x,t)} [\eta_k(x,t,\epsilon) - \frac{1}{2}(B_{2k}^2(x,t)+B_{2k-1}^2(x,t))]$.

The formal similarity between formulas (7.6) and (7.1) is far from accidental. Indeed let me confine my attention to a neighborhood of a *fixed* point (x,t) in spacetime given by $\{(x+\epsilon x',t+\epsilon t'): x,t \text{ fixed } |x'|,|t'| \ll \frac{1}{\epsilon}\}$. The solution $u(x+\epsilon x',t+\epsilon t',\epsilon)$ satisfies equation (2.1) in the stretched variables (x',t'). I will show that as $\epsilon \longrightarrow 0$ it agrees asymptotically with a quasiperiodic m-soliton solution of (2.1) given by relations (2.12-13) with $B_0^2, B_1^2, \ldots, B_{2m}^2$ playing the role of $\xi_0, \xi_1^+, \xi_1^-, \xi_2^+, \ldots, \xi_{2m}^-$. Indeed I can show that the value η_k oscillates in the interval $[B_{2k}^2, B_{2k-1}^2]$ and satisfies the multisoliton evolution equations:

(7.7a) $\qquad\qquad \dfrac{\partial \eta_k}{\partial x'} = \dfrac{2\tilde{\mathcal{R}}(\eta_k)^{1/2}}{\displaystyle\prod_{\substack{i=1 \\ i \neq k}}^{m} (\eta_k - \eta_i)}$,

(7.7b) $\quad \dfrac{\partial \eta_k}{\partial t'} = 2\{-B_0^2 - 2\eta_k + 2 \sum_{i=1}^{m} [\eta_k - \frac{1}{2}(B_{2i}^2 + B_{2i-1}^2)]\} \dfrac{\partial \eta_k}{\partial x'}$,

where $k = 1,2,\ldots,m$ and

(7.7c) $\qquad\qquad \tilde{\mathcal{R}}(\eta) = \displaystyle\prod_{i=0}^{2m} (\eta - B_i^2)$.

The implication is that there are two representations of the solution $u(x,t,\epsilon)$ each having the form of relations (2.12-13). On the one hand is the original global representation which is uniform in (x,t) but involves $O\left(\frac{1}{\epsilon}\right)$ wave parameters given by the spectral microstructure of the operator $\epsilon^2\dfrac{d^2}{dx^2} + v(x)$. On the other hand the local representation (7.6-7) is valid only in a neighborhood of a fixed point

331

(x,t) but describes the solution in terms of only a handful of wave parameters namely $\beta_0^2,...,\beta_{2m}^2$ whose number and values depend on x and t and is independent of ϵ. It is natural to call the intervals $(\beta_{2k-1}^2, \beta_{2k-2}^2)$ effective bands and the intervals $(\beta_{2k}^2, \beta_{2k-1}^2)$ effective gaps where $k = 1,2,...,m$. The oscillation of the η_k's in the effective gaps has wave numbers and frequencies of order $\frac{1}{\epsilon}$ in the original (unstretched) variables x,t.

Equations (7.7) for the η_k's are derived by recalling that the value of η_k equals the value of the unique unconstrained ν_i in the kth effective gap and therefore satisfies equations (2.13) corresponding to the uniform representation. These equations reduce to (7.7) after extensive cancelations on the right hand side. These cancelations result from our knowledge of:

(i)   the values of all ν_i's other than the η_k's in each effective gap (region r ≠ 0) with exponential accuracy.

(ii)   the averaging assumption on the values of the ν_i's in the effective bands.

The mechanism by which the unconstrained ν_i--or equivalently the value of η_k--passes from one gap of the spectral microstructure to an adjacent one is shown in the following figure:

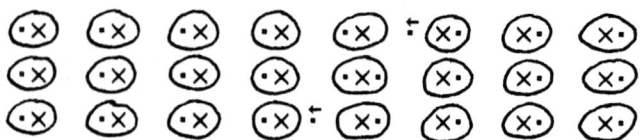

Fig. 7.2

8.   Concluding Remarks

The recapturing of the wave form $u(x,t,\epsilon)$ establishes the connection between the Lax–Levermore approach and modulation theory. The β_k^2's of the Lax–Levermore theory are established as the modulation parameters in the theory of multiphase averaging of Flaschka Forest and McLaughlin [7]. Furthermore the present theory shows, through the evolution equation (6.25), how to calculate beyond the generation of a new phase.

The recapturing of the wave form in the present theory rests heavily on the averaging assumption in the effective bands. This step may be thought of as equivalent to averaging a Lagrangian in Whitham's [20] version of modulation theory, or averaging a system of conservation laws in the Flaschka Forest and McLaughlin [7] version. A rigorous proof of the averaging assumption seems to need a nontrivial extension of this work.

I will finish with a final note on the rate of generation of new oscillatory phases. In the case of periodic initial data a calculation which I have performed in collaboration with David Levermore shows that as t increases the number of phases generated grows linearly in t. Of course, as $\epsilon \longrightarrow 0$ all the results I have described are valid only when $t \ll \frac{1}{\epsilon}$. As t becomes comparable to $\frac{1}{\epsilon}$ the effective local spectral structure of the solution, i.e., the β_k^2's, tends to become as fine-grated as the underlying original microstructure. The theory no longer gives a simple local representation of the solution. One only knows that due to the almost periodicity in time of the solution of the kdV equation a close return to the original data will be made. This must occur after the breakdown of the effective structure. Therefore it must occur only after a time of order $O\left[\frac{1}{\epsilon}\right]$ has elapsed.

References

[1]   V.S. Buslaev and V.N. Fomin, "An inverse scattering problem for the one-dimensional Schrödinger equation on the entire axis," Vestnik Leningrad Univ. 17, 1962, 56–64 (In Russian).

[2]   A. Cohen, T. Kappeler, "Scattering and Inverse Scattering for Steplike Potentials in the Schrödinger Equation," Indiana U. Math J., Vol 34, #1, 1985, 127–180.

[3]   P. Deift, E. Trubowitz, "Inverse Scattering on the Line," Comm. Pure Appl. Math. 32, 1979, 121–252.

[4]   B.A. Dubrovin, V.B. Matveev, and S.P. Novikov, "Nonlinear equations of Korteweg-de Vries type, finite zoned linear operators, and Abelian varieties," Uspekhi Mat. Nauk. 31, 1976, 55–136.

[5]   F.J. Dyson, "Old and New Approaches to the Inverse Scattering Problem," Studies in Math Physics, Princeton Series in Physics (Lieb, Simon, Wightman eds.) 1976.

[6]   L. Faddeev, "The Inverse Problem in the Quantum Theory of Scattering," J. Math. Phys., Vol 4, #1, Jan. 1963, 72–104.

[7]   H. Flaschka, M.G. Forest and D.W. McLaughlin, "Multiphase Averaging and the Inverse Spectral Solution of the Korteweg-de Vries Equation," Comm. Pure Appl. Math. 33, 1980, 739–784.

[8]   C.S. Gardner, J.M. Green, M.D. Kruskal, R.M. Miura, "Method for solving the Korteweg-de Vries equation," Phys. Rev. Lett. 19, (1967), 1095–1097.

[9]   I.M. Gelfand, B.M. Levitan, "On the determination of a differential equation from its spectral function," Izv, Akad.

Nauk SSR. Ser. Math., 15 (309-60). Eng. Translation: Am. Math. Soc. Translation (2), 1, 253 (1955).

[10] I. Kay, H.E. Moses, "Reflectionless Transmission through Dielectrics and Scattering Potentials," J. Appl. Phys. 27, 1956. 1503-1508.

[11] P.D. Lax, "Integrals of Nonlinear Equations of Evolution and Solitary Waves," Comm. Pure Appl. Math., Vol 21 (1968), 467-490.

[12] P.D. Lax and C.D. Levermore, "The Small Dispersion Limit of the Korteweg-de Vries Equation," I, II, III Comm. Pure Appl. Math. 36, 1983, 253-290, 571-593, 809-829.

[13] H.P. McKean and E. Trubowitz, "Hill's operator and hyperelliptic function theory in the presence of infinitely many branch points," Comm. Pure Appl. Math. 29, 1976, 146-226.

[14] H.P. McKean and P. vanMoerbeke, "The spectrum of Hill's equation," Invent. Math. 30, 1975. 217-274.

[15] H.P. McKean, "Partial Differential Equations and Geometry," Proc. Park City Conference, editor C.I. Byrnes, Marcel Dekker, Inc., New York, 1979, 237-252.

[16] E. Trubowitz, "The inverse problem for periodic potentials," Comm. Pure Appl. Math. 30, 1977, 321-337.

[17] S. Venakides, "The zero dispersion limit of the Korteweg-de Vries equation with non-trivial reflection coefficient," Comm. Pure Appl. Math. 38, 1985, 125-155.

[18] S. Venakides, "The generation of modulated wavetrains in the solution of the Korteweg-de Vries equation," Comm. Pure

Appl. Math. 38, 1985, 883-909.

[19]   S. Venakides, "The zero dispersion limit of the periodic kdV
         equation," AMS Transactions in press.

[20]   G.   Whitham,   "Linear   and   Nonlinear   Waves,"   Wiley
         Interscience, New York, 1974.